Graphical Communication

Book 1

A YARWOOD

Chief Examiner in Technical Drawing for
the West Midlands Examinations Board

Chief Examiner in Technical Drawing for a
G.C.E. Examinations Board at Ordinary Level

Nelson

Thomas Nelson and Sons Ltd
Nelson House Mayfield Road
Walton-on-Thames Surrey KT12 5PL

P.O. Box 18123 Nairobi Kenya

116-D JTC Factory Building
Lorong 3 Geylang Square Singapore 1438

Thomas Nelson Australia Pty Ltd
19–39 Jeffcott Street West Melbourne Victoria 3003

Nelson Canada Ltd
1120 Birchmount Road Scarborough Ontario M1K 5G4

Thomas Nelson (Hong Kong) Ltd
Watson Estate Block A 13 Floor
Watson Road Causeway Bay Hong Kong

Thomas Nelson (Nigeria) Ltd
8 Ilupeju Bypass PMB 21303 Ikeja Lagos

© A. Yarwood 1979
First published 1979
Reprinted 1980, 1981, 1982
ISBN 0-17-431260-1
NCN 280-3112-3

Designed by: SGS Associates (Education) Ltd
8 New Row London WC2

By the same author

Technical Drawing Books 1–3
    A course to CSE and GCE O Level

An Introduction to Technical Drawing Books 1 and 2

Woodwork and Design Books 1 and 2

Technical Drawing Graphics

Phototypeset by Tradespools Ltd Frome Somerset
Printed and bound in Hong Kong

Contents

Preface **5**
Equipment **6**
Colour **10**
Papers **12**
Geometrical Construction **14**
    Division of Lines **17**
    Angles **18**
    Triangles **20**
    Polygons **21**
    Circles **25**
    Tangents **26**
    Circles to Triangles **28**
    Similar Polygons (Area) **30**
    Calculating Areas **30**
Logograms **33**
Ideograms **34**
EITB Standard Symbols **35**
Protractor Practice **36**
Optical Illusions **37**
Graphs and Charts **38**
Golden Mean Proportions **44**
Geometrical Patterns **44**
Thread Patterns **48**
Building Drawing **49**
Vectors **54**
Electrical Circuits **58**

Methods of Graphical Illustration **61**
    Isometric Drawing **62**
    Cabinet Drawing **67**
    Planometric Drawing **68**
Orthographic Projection **70**
Symbols and Conventions **71**
Freehand Orthographic Drawing **72**
First Angle Orthographic Projection **74**
    Television Set **75**
    Dowelling Jig **76**
    Disc Brake Pad **77**
    Coat Peg **78**
Third Angle Orthographic Projection **79**
    Tooth Brush Rack **80**
    Spacing Bracket **81**
    Transistor Set **82**
    Electronic Calculator **83**
    Cassette Player **84**
Sections **85**
    Book Trough **86**
    Brake Shoe **87**
    Pipe Bending Jig **88**
Screwed Fastenings **89**
    Film Drying Clip **90**
    Retort Clamp **92**
    Lens Holder **93**
    Cross-feed Mechanism **95**
    Duplicator Crank Arm **96**

Colour pages

Crayon and Water Colour Tinting   between pages **32** and **33**
Thread Patterns   opposite page **48**
Garage Extension Drawing   opposite page **49**

Acknowledgements

The author wishes to place on record his appreciation to the organizations listed below for granting permission to reproduce copyright material in this book.

Examinations Boards

The East Anglian Examinations Board
The East Midland Regional Examinations Board
The South East Examinations Board
The Associated Examinining Board for the General Certificate of Education
The Joint Matriculation Board
The Schools Examinations Department, University of London
The Oxford and Cambridge Schools Examination Board
The Southern Universities Joint Board

The Boards named above have kindly granted permission to reproduce questions from examination papers set in past years.

Photographs

British Thornton Limited have kindly provided four photographs reproduced on pages 6, 7, 8 and 9.

Symbols

The Engineering Industries Training Board have kindly granted permission to reproduce the Standard Symbols on page 35.
Siemens Limited – symbols on page 34.

Logograms and ideograms

Those on page 33 and the colour page facing page 33 are reproduced by kind permission of:
British Rail
The General Electric Company Ltd
Municipal Mutual Insurance Ltd
Independent Television News Ltd
British and Irish Steam Packet Company Ltd
Dupont Ltd
Roadline U.K. Ltd
Yorkshire Bank

Preface

This is the first of two volumes which have been designed as a two-year course for pupils and students who are preparing for examinations in Graphical Communication at the age of 16 +. They are also suitable as textbooks for Technical Drawing examinations. The contents of the books have been compiled from the experience resulting from the author's participation in the following educational activities:

As Head of Craft, Design and Technology departments in Secondary, Grammar and Comprehensive schools;

As the Chief Examiner in Technical Drawing with a C.S.E. examinations board and also with a G.C.E. examinations board and the membership of syllabus revision groups at C.S.E., Ordinary level and Advanced level resulting from these appointments;

As a member of the Crafts, Applied Science and Technology (CAST) subjects committee of the Schools Council; as a member of working parties of that committee; as a CAST member of the 18 + Steering Group for Technical Drawing;

As a member of the working party responsible for PD 308 (Engineering drawing practice for schools and colleges).

Five areas of study have been recognized as being essential to courses in the subject:

Draughtsmanship
Geometry
Orthographic projection
Methods of technical illustration
Technical graphics as a design 'tool'

The acquisition of a good standard of draughtsmanship is necessary for effective graphical communication. Much of the value of technical graphics is lost without good standards of draughting. The disciplines involved in gaining draughting skill are of value. The acquisition of draughting skill cannot, however, be regarded as being of educational importance in itself. The intelligent application of a skill is of far greater educational value than the learning of the skill.

Graphical communication is based in geometry.

A knowledge of constructional geometry is therefore necessary to success in the subject. The fundamental geometrical constructions are described in these books. Exercises are added to emphasize the application of geometry to the shape and form of objects in everyday use. A graphical method of explaining geometrical constructions has been adopted rather than the more common explanations in words.

Orthographic projection is the most important of the methods employed for the full graphical description of three-dimensional artefacts and objects. Students are advised to learn and practice the methods of First Angle and Third Angle projections. A large number of exercises in both First and Third Angle are included.

Many methods of technical illustration are employed in commercial, industrial and media graphics. In this pair of books the methods of technical illustration described have been restricted to isometric drawing, cabinet drawing, planometric drawing and estimated single point and two point perspective drawing. Instrument and freehand methods are given equal value.

Ideas for designs are frequently conveyed by graphical illustration. The student should practice selecting the graphical method best suited to the design concepts he wishes to convey.

Most Technical Drawing books are confined to a specialized form of drawing such as engineering drawing. Many examples of engineering drawing are included within the pages of these two books. Graphical communication has, however, been extended to include applications to building plans and layouts, symbols, geometrical patterns, letter faces, graphs and charts, logograms and ideograms, vectors, electrical circuit diagrams and the making of drawings such as optical illusions and thread patterns. Suggestions for the addition of colour work to drawings are included.

The author deprecates restrictions on the use of equipment such as curve aids, stencils and templates in Technical Graphics. It is hoped the student will be allowed full use of as many drawing aids as is possible. The production of drawings without such aids should, however, be encouraged before aids are made available.

Equipment

To achieve good standards of graphical presentation, good quality drawing equipment is necessary. Good equipment, properly stored and regularly cleaned and serviced, will last for many years.

Drawing boards

Drawing boards are most often made from wood. Boards made from plastics may also be purchased. A common size of board measuring 650 mm by 470 mm is designed for A2 size drawing paper. A3 boards measuring 470 mm by 340 mm are also common in schools. The surfaces of drawing boards should be cleaned by wiping with paper or rags before use. Their surfaces may require sanding from time to time to renew the original flat, smooth and clean surfaces. It is important that drawing board edges are true, and these may require planing straight and square after several years of continuous usage. Sheets of drawing paper may be fixed to drawing boards with pins, clips or self-adhesive tapes such as masking tape or Sellotape. Clips or

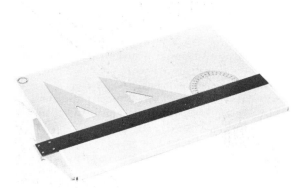

An A2 drawing board, Tee square, set squares and protractor

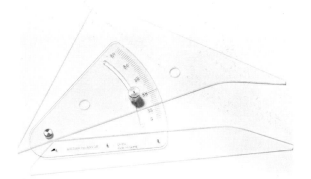

An adjustable set square

tapes are probably best because they do not damage the board's surfaces as do pins.

Tee squares

Tee squares are frequently made from hardwoods, the blades being glued and screwed to the heads. Tee squares made from plastics are becoming increasingly common. It is advisable to use a Tee square of a size suitable to the drawing board to which the drawing sheet has been attached. Tee squares should be regularly cleaned with paper, a clean handkerchief or a duster. The edges of the blades may occasionally need to be planed straight. Tee squares must be stored in such a way as to avoid damage to the blade edges. Screws should be tightened occasionally.

Set squares

Two set squares – 60°, 30° and 45°, 45° – will allow lines to be drawn at a large variety of angles to lines drawn with the Tee square. Not only can 30°, 45°, 60° and 90° lines be drawn, but any angle which is a multiple of 15° can be set up using both 'squares' together. Set squares are best made from transparent plastic materials. Their surfaces should be regularly cleaned by wiping with paper or with a clean handkerchief and occasionally by washing with soap and water. Avoid damaging their straight edges. An adjustable square (photo below) may be preferred.

Protractor

A protractor is an essential item of equipment for setting up angles not easily drawn with the aid of set squares. If an adjustable set square is available, a protractor is not necessary.

Compasses and dividers

A photograph opposite shows a set comprising two 'spring bow' pencil compasses, a 'beam' pencil compass and two pairs of dividers. Such a set will enable the student to draw any circle or arc of radii from about 1 mm up to approximately 250 mm. The dividers are for stepping off accurate measurements on to a drawing. The steel points of compasses and dividers may need to be sharpened occasionally. A small carborundum oil-stone is suitable for this purpose. Pencil leads in compasses should be sharpened on a piece of fine sandpaper or with a fine flat file. It is advisable to use a grade of pencil lead in compasses a little softer than the pencils used for other drawing work. Thus HB or H grade leads in compasses are preferable when 2H or 3H pencils are used for the other pencil work.

A compass set

Erasers

An essential piece of equipment for erasing mistakes is a pencil 'rubber' or eraser. Vinyl erasers are possibly preferable to 'rubbers', but tend to be more expensive. Rubber 'dust' can be a source of annoyance on a drawing sheet and can cause instruments to become soiled. If erasing has been necessary, the eraser waste should be blown or wiped clear before it soils the drawing or the instruments.

Pencils

Most graphical work by students will be in the form of pencil drawings. Seventeen grades of pencil leads are manufactured. Nine grades of 'hard' – 9H to H, followed by F and HB, then six grades of 'black' – B to 6B. The student should adopt the use of two grades of pencil, one for drawing against instruments – say 2H or 3H – the second for freehand drawing – say B or HB. Other grades may be preferred.

Sharpening pencils

After removing some of the wood with a knife, wood chisel or pencil sharpener, the projecting lead may be shaped on a strip of sandpaper or with a smooth flat file, to a sharp round point or to a chisel edge. The student should adopt the method of sharpening which is found to be most suited to his or her methods of drawing – either round point or chisel edge.

Some students may prefer a clutch pencil. This is a device made from metal and shaped like a pencil. It is designed to hold lengths of pencil lead firmly until released by a clutch button at the upper end of the device. Some clutch pencils are fitted with a built-in sharpener for rounding the working end of the lead. Others are made to hold rectangular leads which produce lines drawn to specific widths.

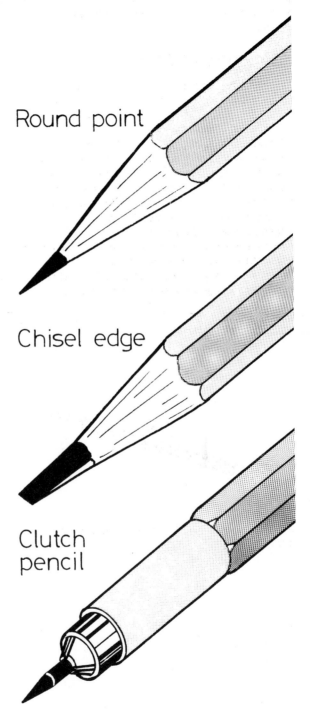

Round point

Chisel edge

Clutch pencil

Parallel motion

Some schools and colleges are equipped with parallel motion devices fitted to drawing boards. These replace Tee squares and drawing boards. The parallel motion straightedge can be slid up and down the board to which it is attached, always remaining parallel to the board top and bottom edges. Horizontal lines are drawn along the straightedge of the motion device. Lines at angles to the horizontal are drawn with the aid of set squares placed along the straightedge.

Drawing machines

Another form of device which can be fitted to a drawing board is the so-called drawing, or draughting 'machine'. This consists of a series of sprung links which enable a hand-held knob to be placed in any position on the surface of the drawing board. A pair of plastic straightedges are attached to the knob. These are set at right angles to each other, but can be locked at any required angle to the board's edges. Tee squares and set squares are not needed with a drawing machine. The hand knob can be locked to its pair of straightedges at any of the angles which can be obtained with Tee and set squares.

An A2 drawing unit with parallel motion

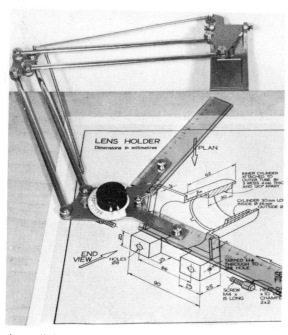

A small draughting machine attached to an A2 drawing board

A variety of curve aids – isometric ellipse template, Flexicurve, French curves and a radius curve.

Curve aids

A large variety of aids, or templates, can be purchased to assist the drawing of smooth and accurate curved lines on drawings. Those most likely to be of interest to the student are listed below.

Radius curves   Made from transparent plastic materials, these consist of a number of different sizes of circular pieces set around the main body of the template. Arcs of from 2 or 3 mm up to about 20 or 25 mm may be easily and accurately drawn with this aid. Of particular value when drawing 'fillets' on an engineering drawing. At the radius centre of each of the circular pieces is a tiny hole through which the arc centre can be located if necessary.

French curves   These are shaped pieces of transparent plastic sheet manufactured in a very large range of different shapes and sizes. A selection of three or four French curves will assist the student in drawing smooth and accurate outlines of any of the curves occurring in his or her drawings.

Ellipse templates   Made from plastic sheet, ellipse templates assist in the speedy, accurate drawing of elliptical outlines. Templates for drawing a large variety of ellipses on different axes are available. One of the most commonly used types of ellipse template is that for drawing isometric ellipses where the axes are in the ratio 1:1.7.

Nut and bolt templates   These enable a draughtsman to draw speedily the various views of hexagonal nuts and bolts. Without this aid, the drawing of such screwed parts can become a very tedious and time consuming chore.

Electrical and electronics templates   These are made for drawing electrical and electronics symbols in circuit diagrams. They consist of a number of symbol outlines cut in thin plastic sheet.

Flexicurves   A tough, rubbery plastic material around an inner flexible core enables these curve aids to be bent to any shape. Once set to the required curve its outline can be drawn along the edge of the Flexicurve to produce a smooth line.

Drawing a curved line

The drawing on this page shows the two stages of drawing a smooth curve with the aid of a French curve. The curved line could also be drawn with the aid of a Flexicurve if one is available.

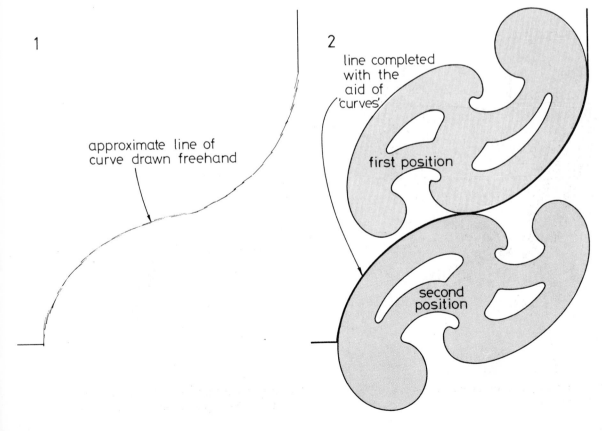

1

approximate line of
curve drawn freehand

2

line completed
with the
aid of
'curves'

first position

second
position

Colour

Colour can be used to advantage to emphasize outline, shape or position of parts of drawings made for graphical communication.

Some rules for colour work

1   Colour should not be included in graphical communication drawings merely for the sake of using colour, but only if drawings are made clearer by the inclusion.

2   When colour is added to drawings it must be applied with care and discretion.

3   Colour washes should be applied to produce thin, even and pale tinting of the areas of drawings to which attention is to be drawn.

4   The application of areas of thick, dark and bright colours which obscure and hide the drawing lines must be avoided.

5   Most clear colours are suitable – reds, blues, yellows, greens and oranges being those most commonly used.

Some uses for colour

1   Tinting parts of engineering drawings such as sectional views.

2   Tinting and outlining parts of building drawings to emphasize materials such as brickwork, woodwork, areas of concrete, surfacing of paths.

3   Tinting and outlining parts of drawings which show additions to existing work.

4   Colouring lines and symbols in drawings such as electrical circuits to show the paths of specific parts of the circuit.

5   Colouring parts of flow diagrams to draw attention to particular details.

6   Tinting, shading or outlining design drawings.

7   Colouring graphics such as logograms, ideograms and symbols.

8   Coloured lines in geometrical constructions to show particular forms of construction or to emphasize the end result of a construction.

9   Outlining or tinting new views.

10   Tinting and outlining pictorial drawings.

11   Shading by the application of different depths of tinting in pictorial drawings.

12   Tinting backgrounds to all forms of graphical communication drawings.

Note   Although a fairly comprehensive list of examples where colour can be used is given above, it should be noted that many graphical communication drawings do not need the addition of

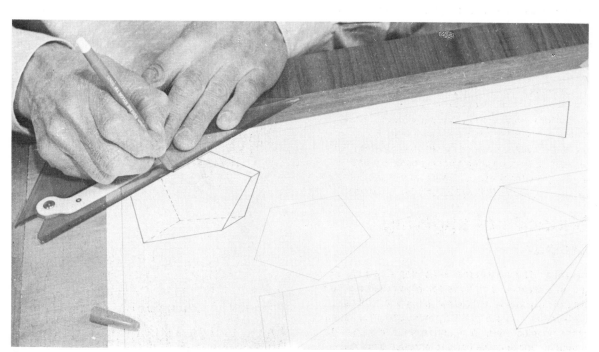

Outline colouring parts of a 'solid' geometry problem

Colour washing with a water colour felt pen

on a drawing. Wax crayons are suitable for hard, clear outline lines, but the wax tends to rub off on to other drawings. Crayon colouring should be 'fixed' by spraying the drawing with a clear, thin fixative varnish from an aerosol.

Pens   Water colour felt pens provide an easy method of applying colour washes over large surfaces of drawings. Nylon tip and ball-point pens are more suitable for drawing hard, clean and bright lines.
Technical pens of the 'Rotring' type will be referred to in Book 2.
Felt tip washes tend to fade in time, particularly in bright sunlight.

Water colour   Thin water colour washes, applied with a painting brush, provide the best and most long-lasting method of colouring drawings. Such washes, carefully applied, enhance the appearance of a drawing and so provide a good graphical impact. The method of application is:

1   'Paint' the area to be colour washed with clean, uncoloured water.
2   Mix a pale wash in a palette.
3   Apply the wash with a brush, working from top to bottom or from side to side.
4   Remove surplus colour with a dry brush.

colour. As stated above under **Some rules for colour work** – Colour should be applied with care and discretion and only if its use enhances the value of a drawing as a means of graphical communication.

Colouring mediums

Four mediums are suggested for colouring drawings:

1   **Crayons**   Pencil crayons (*caran d'ache* type); pastel crayons; wax crayons
2   **Various pens**   Felt tip pens (painting sticks); nylon tip pens; coloured ball-point pens
3   **Water colours**   Water colour in block form; tubes of water colour; powder colour (poster paints)
4   **Dry transfer sheet**   Letrafilm in various colours – rather expensive (Follow maker's instructions.)

Methods of applying colour

Crayons   The easiest method of applying a colour tint on a drawing is with the common *caran d'ache* type of crayon pencil. When it is used to apply a surface tint, a 'grain' will be produced by the action of the pencil strokes. When sharpened to a fine point, pencil crayons can be used to produce coloured outlines. Pastel crayons tend to leave a dusty surface

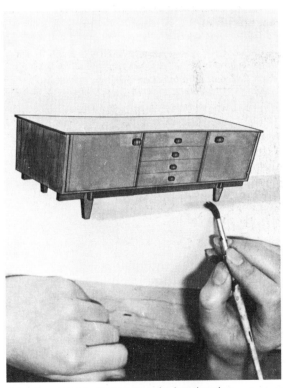

Applying colour washes to a design drawing

Papers

There is a vast range of different papers made for a variety of purposes. Nearly all paper is made from the vegetable fibres obtained from wood, straw, esparto grass, cotton and rag wastes. The cheapest papers, such as those on which newspapers are printed, are manufactured almost entirely from wood. Better quality papers will contain cotton fibres. It is the cellulose of these vegetable fibres which forms the bulk of paper. Lignin, which is the bonding material holding together the cellulose in vegetable fibres, is an important constituent of paper. Other materials are used in paper making, but to a far less extent than are these vegetable fibres.

A common method of measuring paper is by its weight at so many grams per square metre – gsm or g/m^2. An A0 sheet of paper is 1 square metre.

Drawing papers

A variety of papers are used on which drawings for graphical communication may be made. Notes on the more common of these are given below.

Cartridge paper This is the paper the student is most likely to use for most of his drawings. So called because this paper was originally used for making cartridge shell cases. Made mainly from esparto grass in a variety of thicknesses. That most frequently used in schools and colleges is 90 gsm but lighter (and heavier) cartridge papers are made. Cartridge paper is a good quality drawing paper which takes pencil and colour well. Its surfaces will resist a fair amount of rubbing before becoming damaged. This enables the student to erase mistakes with some confidence.

Detail paper A good quality, but thin paper, which is a good medium for pencil and colour work. A rag paper. Usually sold in rolls. Weight – 50 gsm. Can be used as a tracing paper because of its thinness.

Grid papers Usually made from 90 gsm cartridge paper of A4 size, grid papers are printed with a lattice of lines, either at right angles to each other – square grids – or with lines on axes of 30° and 90° – isometric grids. Other grid papers are made, one of which is printed with a pattern of dots at the corners of 10 mm or 20 mm squares. The lines are printed at a variety of distances apart. Thus 2 mm, 5 mm, 10 mm or 20 mm square grid papers can be purchased as can 5 mm, 10 mm and 20 mm isometric papers. Other grids are printed, notably perspective grids. Grid lines are printed blue (or green) because photographic plates used for printing purposes are not sensitive to blues.

Tracing papers 'Natural' tracing paper is a high quality tracing material which can be purchased in 25 metre rolls. Weights of 38 gsm and 63 gsm are common. Other tracing materials are flimsy transparent papers which may be very suitable for transferring a shape from one drawing to another. Polyester tracing films such as 'Permatrace' are expensive and are not usually found in schools. Their great strength and the fact that they do not shrink or expand makes them particularly suitable for some drawing purposes.

Papers for ink work When making drawings in Indian ink, more expensive papers or boards may be required, although good quality ink work can be carried out on good cartridge paper or on natural or plastic tracing sheet. Thicker papers, or boards, with smooth white or ivory surfaces are commonly used for ink work. One rather expensive board of this type is Bristol Board, sold in thicknesses known as 'one-sheet', 'two-sheet', 'three-sheet' and 'four-sheet'.

Sizes of drawing sheets

Schools, colleges and examining boards now almost exclusively use the A series of drawing sheets. The edge lengths of all sheets in this series are in the same proportion of $\sqrt{2}:1$ as shown in the drawing. An A0 sheet is exactly 1 square metre in area, and as an A1 sheet is exactly half an A0 sheet, an A1 sheet is 0·5 ($\frac{1}{2}$) square metre. An A2 sheet is 0·25 ($\frac{1}{4}$) square metre and so on down the series. Larger sheets than A0 are 2A0 (2 square metres) and 4A0 (4 square metres).
The sheets most frequently used by students are A2, A3 and A4.

Dimensions of the A series

2A0	1189 mm × 1682 mm	
A0	841 mm × 1189 mm	(1 square metre)
A1	594 mm × 841 mm	
A2	420 mm × 594 mm	
A3	297 mm × 420 mm	
A4	210 mm × 297 mm	

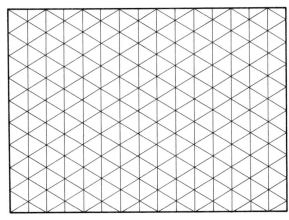

All A size sheets edge proportions are $\sqrt{2}:1$

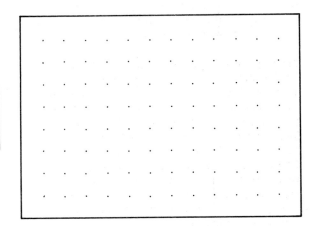

10 mm square dot grid paper

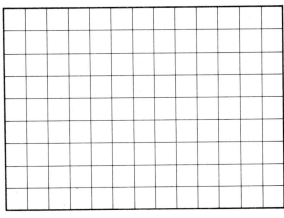

10 mm isometric grid paper

10 mm square grid paper

13

Fastening paper to drawing boards

Drawing sheets can be secured to drawing boards by fastening the sheet corners with pins, by attaching the paper with purpose-made steel clips, or with self-adhesive tapes such as masking tape or Sellotape. If possible avoid pinning paper to drawing boards. The repeated pushing of pins through the paper into the board surfaces makes numerous small holes. When drawing with a pencil on paper over these holes it becomes difficult to draw clean, regular lines.

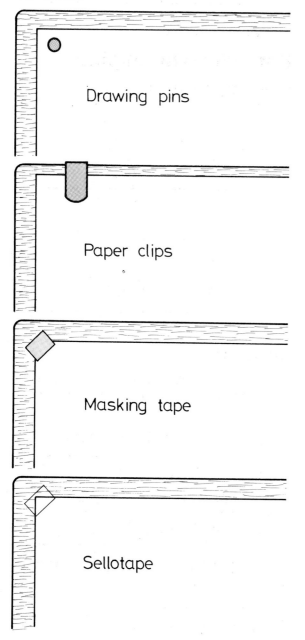

Drawing pins

Paper clips

Masking tape

Sellotape

Geometrical Construction

Graphical communication is based upon a knowledge of geometrical constructions. The following methods of drawing should be employed when working geometrical drawings.

Lines

Outline line – thick and black.

Thin line – thin and black. Dimension lines, chain lines.

Construction lines – very thin. Can be easily erased, but should be clearly visible.

Centre lines – thin and black. A chain line.

Letters and figures

Letters and figures should be well drawn and easy to read.

ABCDEFGHIJKLMNOPQRSTUVWXYZ

abcdefghijklmnopqrstuvwxyz

1234567890

Decimal points

Decimal points should be placed as shown. Figures of less than 1 involving decimals should be preceded by 0.

0.05   0.55   0.9   2.5   6.75

Dimensions

Dimension lines and projection lines should be thin and black. Dimensions should be placed above dimension lines so as to be read from the bottom or from the right-hand side of a drawing. Arrow heads should be about 3 mm long. Larger dimensions should be outside smaller ones. Avoid, if possible, the placing of dimensions within the outline of a drawing.

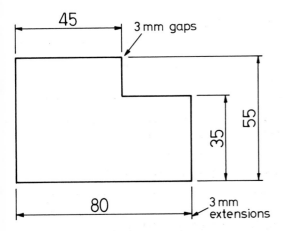

Dimensioning squares

The symbol □ (square) should be placed before the figures of the dimension.

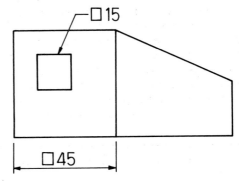

Dimensioning angles

Several methods are shown.

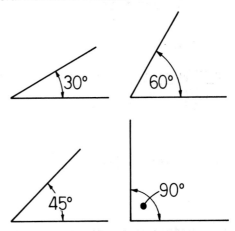

Dimensioning arcs

Lines for dimensioning arcs should pass through, or be in line with, the arc centres. The symbol R (radius) and Ø (diameter) should be placed before the figures of the dimension.

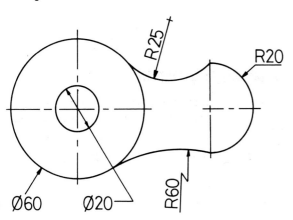

Leaders

Leaders for notes should end in an arrow touching a line on the drawing or in a dot when not touching a line.

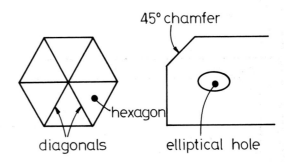

Layout for geometry exercises

Pages 17 to 32 contain descriptions and exercises connected with constructional geometry. Most of the geometrical constructions described in these pages can be worked on A4 size drawing sheets laid out as shown in the drawing on this page. This drawing shows the working of the constructional geometry and Exercise 1 from page 17. The student is advised to follow this form of layout, which produces good technical drawings.

Some of the exercises from pages 17 to 32 can only be worked on sheets of A3 size paper.

How to use pages 17 to 32

Each of the geometrical constructions given on pages 17 to 32 is described in a series of small numbered drawings. These explain the step-by-step procedures needed to complete each of the constructions. Students are advised to follow the drawings, copying each step as in the drawings on this page, in order to learn the constructions. Words are not used to describe the steps. The method employed relies on the drawings. It is thus a 'graphical communications' method.

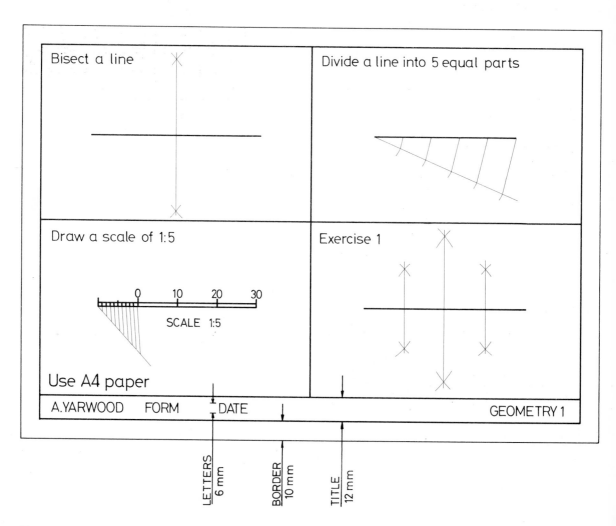

Division of lines

Bisect a line

AB is 87 mm long.
$\simeq$ means approximately equal to.

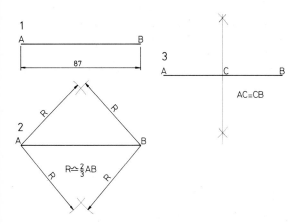

$AC \simeq CB$

$R \simeq \frac{2}{3} AB$

Divide a line into equal parts

AB is 73 mm long.
Divide AB into 5 equal parts.

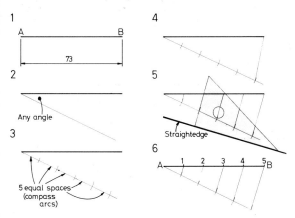

Any angle

5 equal spaces
(compass arcs)

Straightedge

Draw a scale

Scale of 1:5 to read in centimetres (cm) up to 40 cm.

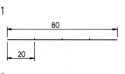

Divide the first 20 mm into 10

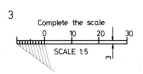

Complete the scale

SCALE 1:5

Exercises

1 Draw a line 83 mm long. Using a geometrical method, divide the line into 4 equal parts. (Answer given on page 16.)

2 Draw a line 112 mm long. Using a geometrical method divide the line into 5 equal parts.

3 Draw a line 94 mm long. Divide the line into seven equal parts using a good geometrical method.

4 Construct a scale of 1:5 to read in cm up to 50 cm.

5 Construct a scale of 1:2 to read in millimetres (mm) up to 240 mm.

6 Fig. 1 is a plan of a TV aerial with eight equally spaced rods to the left of a central support column and ten equally spaced rods to the right. Make a scale 1:1 drawing of the centre lines of the aerial using a good geometrical construction to show the spacing of the rods.

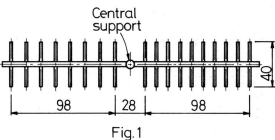

Central support

98   28   98

Fig. 1

Angles

Construct an angle of 60°

AB is 70 mm long.

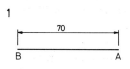

1

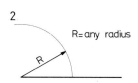

2

R= any radius

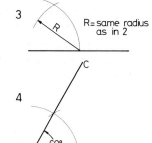

3   R= same radius as in 2

4   60°

Bisect any angle

AB is 65 mm long.

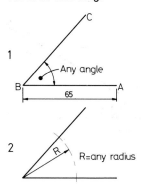

1   Any angle

2   R=any radius

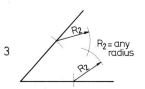

3   R₂= any radius

4   Angle CBD = angle ABD

Construct an angle of 30°

AB is 60 mm long.

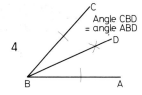

1   60°

2   Bisect 60° = 30°   30°

Construct an angle of 120°

AB is 55 mm long.

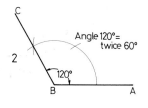

1   60°

2   Angle 120°= twice 60°   120°

Construct an angle of 90°

AB is 65 mm long.

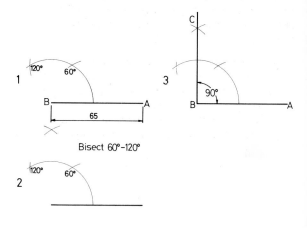

1   120°   60°

Bisect 60°–120°

2   120°   60°

3   90°

Draw an angle with a protractor

Angles which are multiples of 15° may be constructed by bisection. Other angles must be constructed with the aid of a protractor.

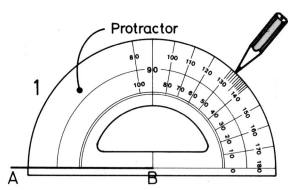

Protractor

1

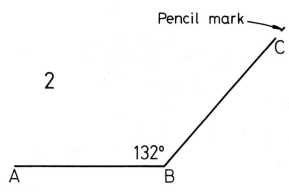

2   Pencil mark   C   132°

Exercises

1   Fig. 1 is a small-scale drawing of a window divided into 15 equal panes of glass. Copy Fig. 1 showing *all* constructions used to obtain the 15 panes.

2   A toy xylophone has eight notes as shown in Fig. 2. The widths of the notes and spaces are equal. Copy and complete the larger, incomplete drawing of the xylophone. Use a geometrical construction to set out the positions of the notes and spaces accurately. (*South East Regional Examinations Board*)

3   Construct the following angles using the geometrical methods shown on page 18: 60°; 30°; 15° (bisect 30°); 90°; 45° (bisect 90°); 120°; 105° (bisect between 90° and 120°); 63° (protractor); 108° (protractor); 145° (protractor).

4   Copy Fig. 3. AB is a lever which can be rotated to AC through eight equally spaced angles. Show by accurate geometrical constructions all the positions of the lever between AB and AC.

5   Fig. 4 shows the face of a central heating temperature gauge. Copy Fig. 4 using compasses and straightedge only.

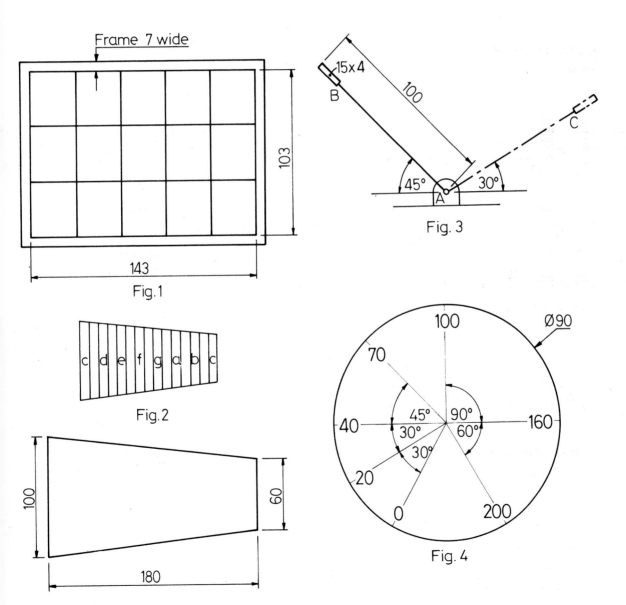

Frame 7 wide

103

143

Fig. 1

c d e f g a b c

Fig. 2

100

180

60

15x4

B

100

45°   30°

A

Fig. 3

100

70

Ø90

40

45°   90°

30°   60°

160

30°

20

0   200

Fig. 4

Triangles

Construct a scalene triangle given three sides

AB = 85 mm; AC = 95 mm; BC = 70 mm.

1

2

3

Construct a scalene triangle given two sides and the included angle

AB = 90 mm; AC = 110 mm; Angle BAC = 42°.

1

2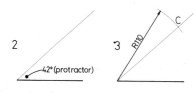

3

4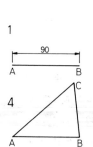

Construct a scalene triangle given two angles and the included side

AB = 60 mm; Angle BAC = 73°; Angle ABC = 54°.

1

2

3

Construct a right-angled triangle

First example AB = 100 mm; BC = 65 mm; Angle ABC = 90°.

Second example AB = 90 mm; AC = 110 mm; Angle ABC = 90°.

Third example AB = 60 mm; BC = 80 mm; CA = 100 mm. This is a triangle with sides in the ratio 3:4:5.

First example

1

2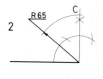

3

Second example

1

2

3
(or with set square)

Third example

1

2

3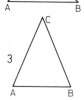

Construct an isosceles triangle

First example AB = 105 mm; AC = BC = 65 mm.

Second example AB = 70 mm; Angle BAC = Angle ABC = 66°.

First example

1

2

3

Second example

1

2
66° (protractor)

3

Construct an equilateral triangle

First example AB = BC = CA = 85 mm.

Second example AB = BC = CA = 90 mm.

First example

1

2

3

Second example

1

2
(or with set square)

3

Polygons

Construct a square

Side lengths are 65 mm.

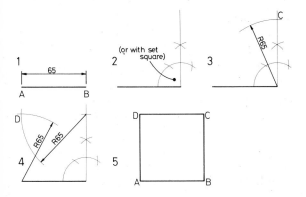

Construct a rhombus

Side lengths are 70 mm; acute angles are 75°.

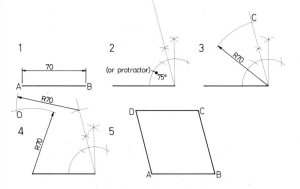

Construct a rectangle

AB = CD = 110 mm; AD = BC = 50 mm.

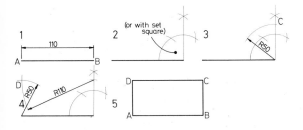

Construct a regular hexagon

First example   Inscribed in circle R35 mm.

Second example   Circumscribing a circle R30 mm.

Third example   With a set square. Side lengths 28 mm.

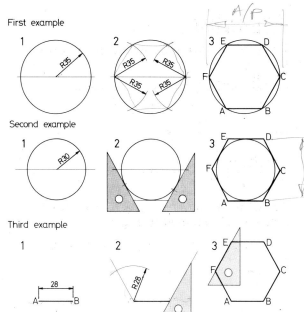

Construct a regular octagon

First example   Within a square of 70 mm sides.

Second example   Within a square of 75 mm sides.

Third example   Inscribed in circle of 80 mm.

Fourth example   Circumscribing a circle of 65 mm.

First example

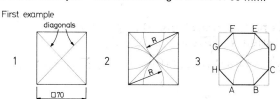

Second example

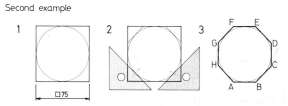

Third example

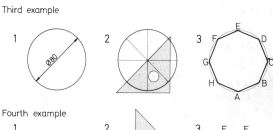

Fourth example

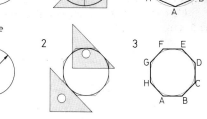

Exercises

1 Construct *all* the following triangles:

Scalene

(a) ABC: AB = 90 mm; BC = 60 mm; CA = 70 mm.

(b) DEF: DE = 112 mm; EF = 94 mm;
FD = 102 mm.

(c) XYZ: XY = 100 mm; XZ = 120 mm; angle
ZXY = 63°.

(d) ABC: AB = 80 mm; BC = 90 mm; angle
ABC = 70°.

(e) STU; ST = 70 mm; angle UST = 68°; angle
STU = 56°.

(f) ABC: AB = 84 mm; angle BAC = 38°; angle
ABC = 43°.

Right-angled

(a) ABC: AB = 110 mm; angle ABC = 90°;
BC = 70 mm.

(b) FGH: angle GFH = 90°; FG = 110 mm;
FH = 80 mm.

(c) JKL: angle JKL = 90°; JK = 76 mm;
JL = 110 mm.

(d) ABC: AB = 83 mm; angle ABC = 90°;
AC = 115 mm.

(e) ABC: AB = 90 mm; BC = 120 mm;
CA = 150 mm.

(f) XYZ: XY = 57 mm; YZ = 95 mm; XZ = 76 mm.

Isosceles

(a) ABC: AB = 95 mm; BC = CA = 83 mm.

(b) XYZ: XY = 76 mm; XZ = YZ = 56 mm.

(c) ABC: AB = 64 mm; angle at A = angle at
B = 73°.

(d) STU: ST = 59 mm; angle at S = angle at
T = 50°.

Equilateral

(a) ABC: AB = BC = CA = 70 mm.

(b) DEF: Each side = 83 mm.

(c) XYZ: XY = 68 mm.

2. Fig. 1 shows a square with a diagonal divided
into parts. Draw Fig. 1 using a suitable construction
for finding E, F and G.

3 Draw the parallelogram ABCD (Fig. 2). Find, and
clearly mark, the centre of AC. All constructions
must be shown.

4 Fig. 3 shows a regular octagon drawn within a
square with its top two corners radiused. Draw Fig. 3
showing all geometrical constructions involved.

5 Fig. 4 is a regular hexagon of 60 mm side lengths
inscribed within a circle. Using compasses and a
straightedge only, construct Fig. 4. Show all
constructions.

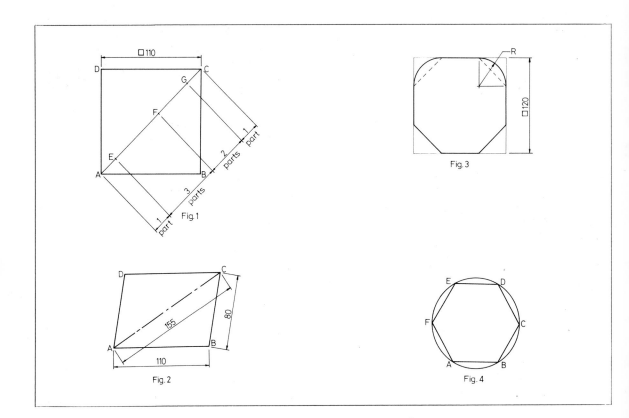

Fig. 1

Fig. 2

Fig. 3

Fig. 4

Polygons, continued

Construct a regular pentagon

First example With the aid of a protractor. Side lengths are 28 mm.

Second example A geometrical method. Side lengths are 30 mm.

First example

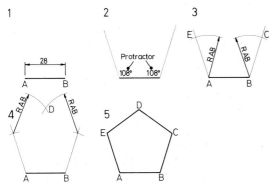

Second example

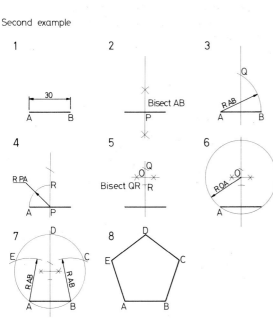

Draw similar polygons of smaller side lengths

Irregular pentagon AB = 80 mm; angle BAE = 90°; angle ABC = 120°; AE = 50 mm; BC = 30 mm; CD = 70 mm; DE = 50 mm. Reduced scale of sides is 3:5 or $\frac{3}{5}$.

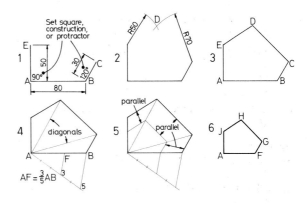

Draw similar polygons of larger side lengths

Irregular hexagon AB = 40 mm; angle BAF = 120°; angle ABC = 135°; AF = 40 mm; BC = 40 mm; angle AFE = 150°; FE = 30 mm; CD = 70 mm; DE = 40 mm.

Increased scale of sides is 5:4 or $1\frac{1}{4}$.

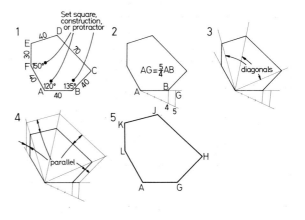

Exercises

1 Fig. 1 shows the plan of a handwheel for a stopcock. The shape is based on a regular pentagon with sides of 30 mm.

Draw, scale 2:1, a plan of the handwheel showing all constructions for obtaining the pentagon and the centres of the arcs. Clearly show all points of tangency.

(East Anglian)

(**Note**   See page 26 for constructions of the tangential arcs.)

2 Fig. 2 is a drawing of an electric meter face. Its outer shape is based on a regular pentagon of 80 mm sides. Make an accurate drawing of the meter face.

3 Fig. 3 is a large scale drawing of a neck pendant. AB should be 20 mm.
Copy Fig. 3 to a scale of 1:1 and reduce your drawing proportionally to obtain the correct size for the pendant.

4 Fig. 4 shows a pentagonal jig. Copy Fig. 4 to a scale of 1:1. Enlarge your drawing proportionally so that the 70 mm side is enlarged to 100 mm. On your enlarged drawing divide AB into three equal spaces.

5 Fig. 5 shows the shape of a guard plate from a lathe. Draw Fig. 5 to a scale of 1:5. Enlarge your drawing so that AB is 1·5 times as long.

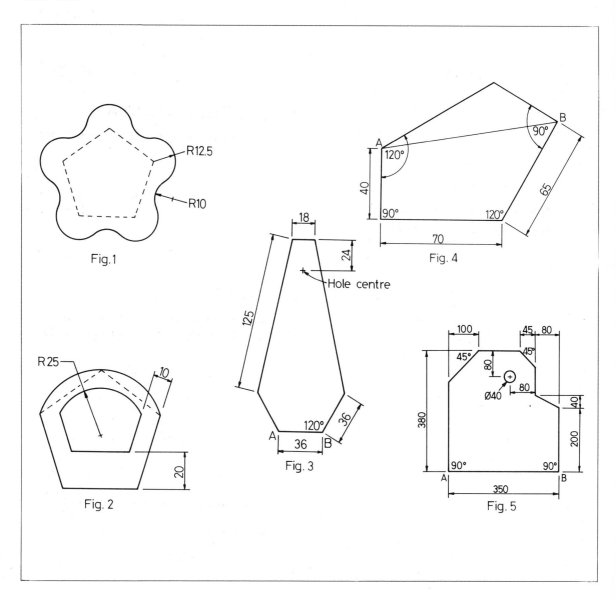

Fig. 1

Fig. 2

Fig. 3

Fig. 4

Fig. 5

Circles

Parts of a circle

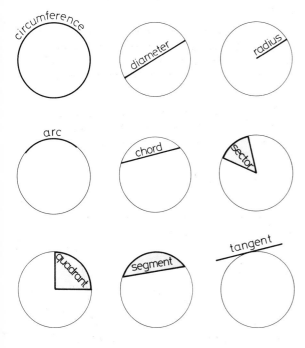

Radius angles

First example   Angle a right angle. R30 arc.

Second example   Any acute angle. R35 arc.

Third example   Any obtuse angle. R40 arc.

First example

R=30

Second example

R=35

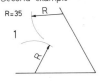

Third example

R=40

Angle within a semi-circle

The angle within a semi-circle is 90°. A 90° angle construction can be based on this geometrical fact.

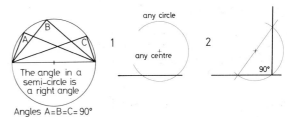

The angle in a semi-circle is a right angle

Angles A=B=C= 90°

Straight line tangents on circles

Draw a circle of any radius. Take any points on the circle, P and Q.

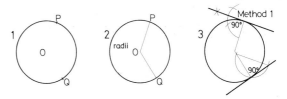

Straight line tangents to circles

Draw any circle of centre O. Take any point P outside the circle.

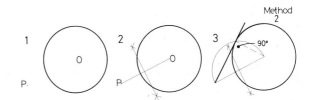

Exercise

Construct the triangle ABC of Fig. 1 making use of the fact that the angle within a semi-circle is 90°.

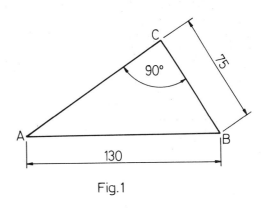

Fig.1

Tangents

External straight line tangent to two circles

Circle O of Ø40 mm. Circle C of Ø70 mm. Draw the external tangent to the two circles.

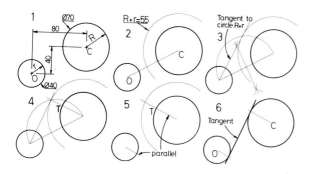

Internal straight line tangent to two circles

Circle O of Ø40 mm. Circle C or Ø70 mm. Draw the internal tangent to the two circles.

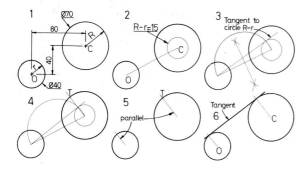

Tangential arcs

First example   Outline of a chain link.

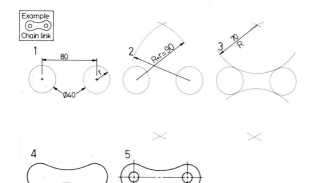

Second example   Outline of a gasket.

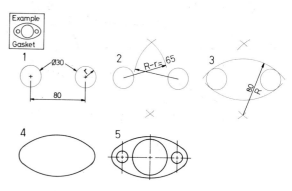

Exercises

1   Draw two circles of diameters 60 mm and 40 mm with their centres 65 mm apart. Construct both common external straight line tangents to the two circles.

2   Draw the same pair of circles as in Exercise 1 and construct both common internal straight line tangents to the two circles.

3   Make an accurate drawing of the file handle shown in Fig. 1 showing the constructions for finding the centres of the tangential arcs.

4   Fig. 2 is a small-scale plan of the end of a stair rail. Copy Fig. 2 to a scale of 1:1.

5   Make a 1:1 scale drawing of the clip shown in Fig. 3.

6   Fig. 4 is a large-scale drawing of the handle from a car ignition/steering-lock key. Make an accurate scale 1:1 drawing of the handle showing all constructions, including those for the straight line tangents from A and B to the arcs centred at C and D.

7   Fig. 5 is a drawing of a tool to fit a special adjusting screw. Make an accurate scale 1:1 drawing of the tool showing all the constructions for the straight line tangents.

8   Copy the outline of the spanner in Fig. 6 showing clearly all tangential construction.

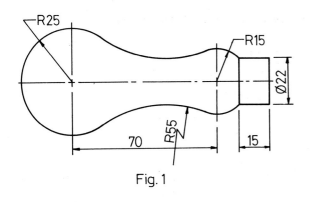

Fig. 1

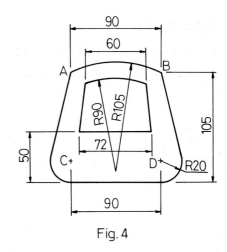

Fig. 4

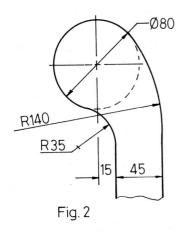

Fig. 2

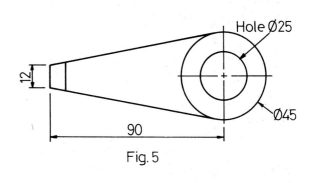

Fig. 5

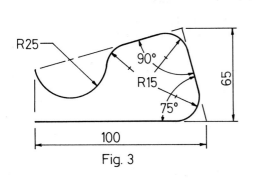

Fig. 3

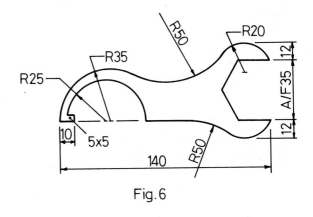

Fig. 6

27

Circles to triangles

Inscribed circle

Triangle of side lengths 100 mm, 80 mm and 116 mm.
Draw the inscribed circle to the triangle.

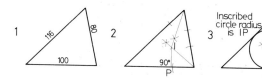

Circumscribed circle

Triangle of sides 80 mm and 100 mm. Included angle 60°.
Draw the circumscribed circle to the triangle.

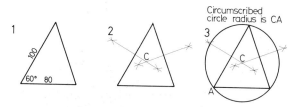

Ellipses

Auxiliary circle method

Major axis 120 mm. Minor axis 70 mm.

Draw ellipse on these axes using the auxiliary circles method.

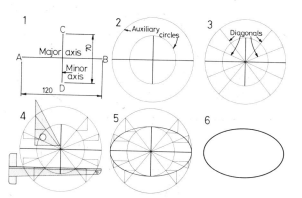

Trammel method

Major axis 120 mm. Minor axis 70 mm.
Draw ellipse on these axes using a trammel.

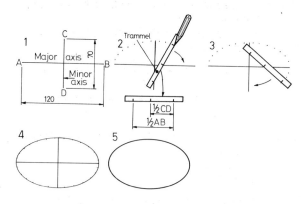

String method of drawing an ellipse

Major axis 86 mm. Minor axis 48 mm.
Draw ellipse using the string method.

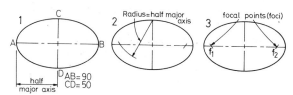

Focal points of an ellipse

Ellipse on axes of 90 mm and 50 mm.
Find the focal points f_1 and f_2 on the major axis.

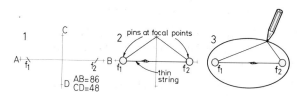

Tangent on an ellipse

Ellipse on axes 100 mm and 56 mm.
Construct a straight line tangent at any point T.

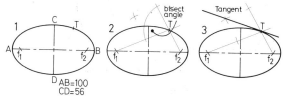

Exercises

1 Construct the parallelogram shown in Fig. 1.
Draw the diagonal AC.
Within triangle ACD construct and draw the inscribed circle. State its diameter.
Construct and draw the circumscribed circle to the triangle ABC. State its diameter.

2 Construct a triangle ABC in which AB = 120 mm; angle ABC = 57°; the vertical height of C above AB = 90 mm. Construct and draw the inscribed circle to the triangle ABC.

3 Construct the triangle PQR. PQ = 85 mm; angle PQR = 65°; angle QPR = 71°.
Construct and draw the circumscribed circle to triangle PQR.

4 Fig. 2 shows an equilateral triangle of sides 80 mm, with its circumscribing circle. Within the triangle are drawn three touching equal circles. Copy Fig. 2 to a scale of 1:1 showing clearly all constructions.

5 Copy the drawing Fig. 3 to a scale of 1:1 showing all the constructions needed to obtain the ellipse.

6 Fig. 4 shows the semi-elliptical end of a cast-iron central heating radiator. Make an accurate copy of Fig. 4 to a scale of 1:1.

7 Construct an ellipse on major and minor axes of 130 mm and 75 mm respectively and foci F and F_1. Locate a point P on its periphery such that FPF_1 is a right angle. At point P construct a tangent.
(Joint Matriculation Board)

8 A light is shone through a hole cut in a screen on the surface of a prism as shown in Fig. 5. The light rays are parallel.
The true shape of the image thrown on to the prism is a 70 mm diameter circle.
Construct the shape of the hole cut in the screen.
(Oxford and Cambridge)

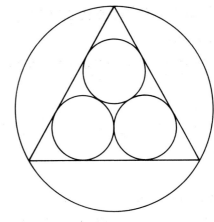

Fig. 2

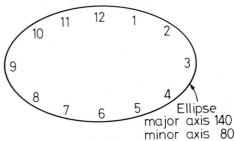

Fig. 3

Ellipse
major axis 140
minor axis 80

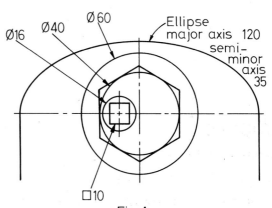

Ø16   Ø40   Ø60   Ellipse major axis 120 semi-minor axis 35

☐10

Fig. 4

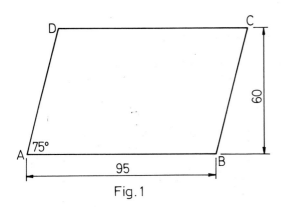

75°

60

95

A   B   C   D

Fig. 1

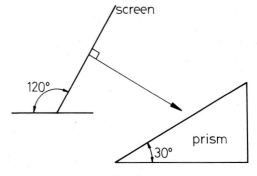

screen

120°

prism

30°

Fig. 5

Similar polygons (area)

Reduce in area

Quadrilateral AB = 50 mm; angle BAD = 80°;
AD = 45 mm; BC = 60 mm; CD = 65 mm.
Reduction in area of 2:3 or $\frac{2}{3}$.

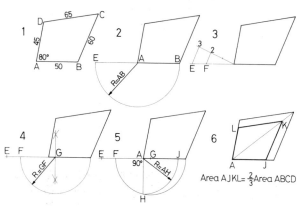

Area AJKL= $\frac{2}{3}$Area ABCD

Enlarge in area

Pentagon AB = 45 mm; angle at A = 110°; angle
at B = 90°; AE = 50 mm; BC = 60 mm;
CD = 20 mm; DE = 50 mm. Enlargement of area of
4:3 or $\frac{4}{3}$.

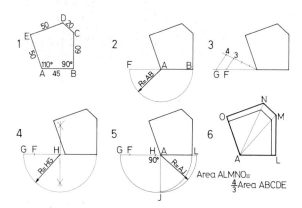

Area ALMNO=
$\frac{4}{3}$Area ABCDE

Calculating areas

Counting squares

Draw the given figure.
Find its area by adding squares.

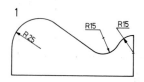

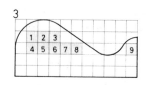

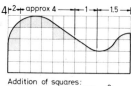

Addition of squares:
11x 2=22+9+2+4+1+1.5 = 39.5 cm²

Mid-ordinate method

Draw the same figure.
Find its area using the mid-ordinate method.

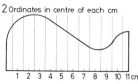

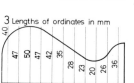

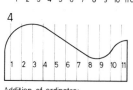

Addition of ordinates:
40+47+50+47+42+35+28+23+20+26+36
=394 mm = 39.4 cm
Area = 39.4 cm²

Exercises

1 Construct a regular pentagon ABCDE of sides
65 mm. Enlarge the pentagon to one with an area
1·5 times that of ABCDE.

2 Construct a regular hexagon ABCDEF of side
lengths 50 mm. Reduce this to a regular hexagon of
area 0·6 that of ABCDEF.

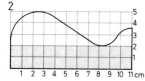

3 Draw the polygon given in Fig. 1 to a scale of 1:1.
Construct a similar polygon with an area 1·25 times
that of the given polygon.

4 Draw Fig. 2 to a scale of 1:1. Construct a similar
polygon of area 0·75 times that of Fig. 2.

5 A section of aerofoil, drawn to a suitable scale,
is given in Fig. 3.
(a) Draw the section.
(b) Use a geometrical construction to determine the
     area of the section and print this dimension in
     square centimetres, accurate to one decimal
     place, under your drawing.

(London)

6 Fig. 4 is a drawing of a sheet of plastic in which
three holes have been cut – an ellipse of major
axis 100 mm and minor axis 65 mm, and two circles,
each of diameter 30 mm.
Draw Fig. 4 to a scale of 1:1 and, using one of the
methods described on page 30, determine the area
of the plastic sheet remaining after the holes have
been cut away. Print the area in square millimetres
next to your drawing.

7 Fig. 5 shows the shape, drawn to a small scale,
of the side of a leather hand case. Copy Fig. 5 to a
scale of 1:1 and determine the area of your drawing
in square centimetres. Print the area beneath your
drawing.

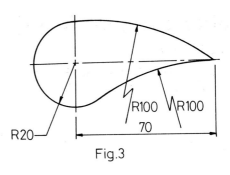

Fig.3

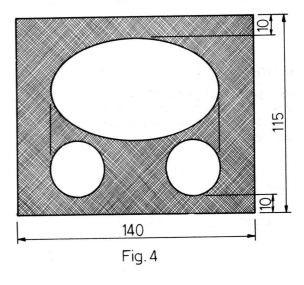

Fig. 4

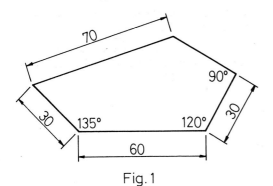

Fig. 1

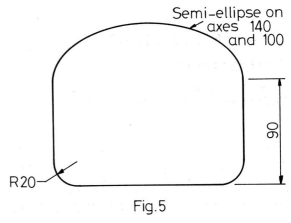

AB = 50
AD = 40
BD = 75
BC = 75
CD = 110

Fig. 2

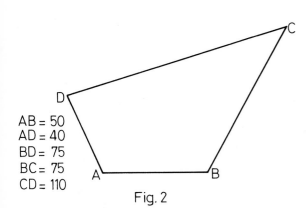

Fig.5

31

Exercises (General)

1 Fig. 1 is a development for the metal coat hook shown in Fig. 2. Draw Fig. 1 to a scale of 1:1 showing all the constructions needed to achieve an accurate outline.

2 A dimensional elevation of a four-bladed outboard engine propellor is given in Fig. 3. Draw to a scale of 1:1 one blade showing all the geometrical construction.
(East Anglian)

3 The bearing of a ship at sea is $337\frac{1}{2}°$ from A and 276° from B as shown in Fig. 4. Later it is sighted again at a bearing of 045° from A and $337\frac{1}{2}°$ from B.
(a) Plot its second position.
(b) State how far the ship has travelled in km.
(c) Show by construction and state the *shortest* distance between the ship and the coast at position 1 in km.
Work to a scale of 10 mm = 1 km.
(South East Regional Examination Board)

4 Draw to a scale of 1:1 the outline of the telephone shown in Fig. 5. All constructions for centres of radii must be shown. Mark clearly each point of tangency.
(East Midland Regional Examinations Board)

5 Fig. 6 gives the plan and dimensions of one of a pair of 'carpenter's clamps'. The small sketch shows how these clamps are self-locking and hold more tightly as the work is pushed into them.
(a) Draw, to a scale of 1:1, the given view, showing clearly the geometrical constructions for finding the centres of the arcs, and the points of tangency of the various arcs.
(b) Assuming that a range of these clamps is to be made, make a plain scale from which measurements could be taken in order to draw a similar plan, in which the pivot hole is O16, i.e. 16 mm represents 10 mm, or a scale of 1·6:1.
(S.U.J.B.)

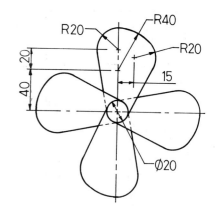

Fig. 3

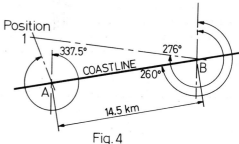

Fig. 4

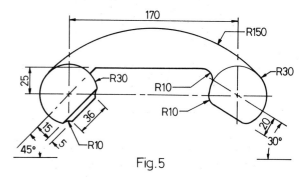

Fig. 5

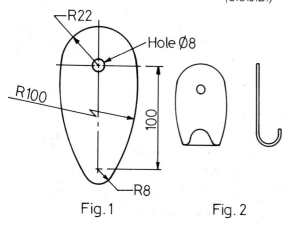

Fig. 1   Fig. 2

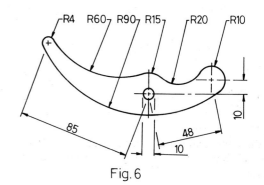

Fig. 6

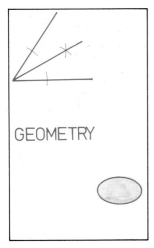

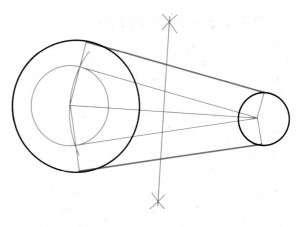

1 The cover for a folder containing drawings of geometrical constructions. Lines and title drawn with a coloured felt tip pen; tinting with a coloured pencil crayon.

4 The geometrical construction for common external tangents to two circles. The construction lines and tangents have been drawn with a coloured ball-point pen.

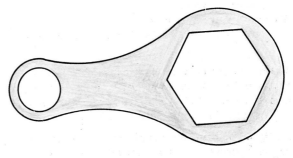

2 A geometrical pattern. Some lines drawn with coloured felt tip pen; tinting with coloured pencil crayon.

5 Drawing of a spanner, tinted with coloured pencil crayoning.

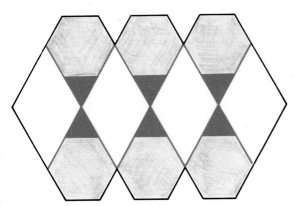

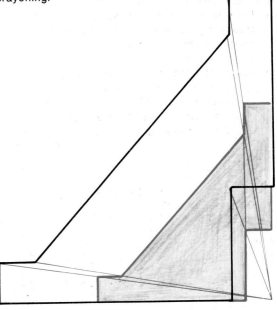

3 Geometrical pattern. Some lines and shading with felt tip pen; other shading with coloured pencil crayon.

6 Geometrical construction for reducing a section through a moulding. Construction and reduced shape drawn with coloured ball-point pen; tinting with coloured pencil crayoning.

Crayon and Water Colour Tinting

Way in

Way out

Morocco

Tunisia

Turkey

Logograms

The drawings on this page are all based on letters of the alphabet. Such graphical drawings are known as 'logograms'.

Exercises

Draw each of the seven logograms to a scale of 1:1 with the aid of drawing instruments. Use coloured crayons, coloured pens or water colour washes to tint the areas of the drawings which are shown shaded.

The dimensions for the Yorkshire Bank logogram can be taken from the freehand drawing on page 73.

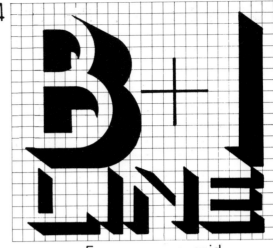

4

5mm square grid

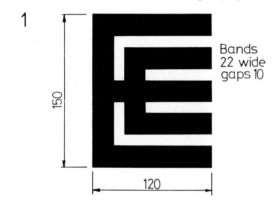

1

Bands
22 wide
gaps 10

150

120

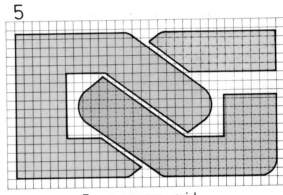

5

5mm square grid

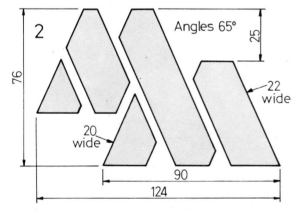

2

Angles 65°

76

25

22 wide

20 wide

90

124

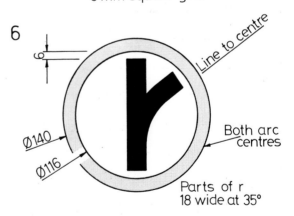

6

6

Line to centre

Ø140

Ø116

Both arc centres

Parts of r
18 wide at 35°

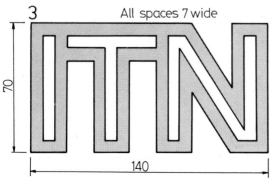

3

All spaces 7 wide

70

140

7

Yorkshire Bank

Ideograms

The symbols on this page all represent appliances in common use. Such symbols are known as 'ideograms'.

Exercises

1 Drawings 1 to 6 show the ideograms on the faces of six switches from the dashboard of a motor car. State in writing which of the car's appliances are controlled by each of the switches.
Different manufacturers employ different ideograms to display on such switches which control the appliances in the vehicle they manufacture.
Make accurate drawings of the ideograms found on control switches fitted to motor cars other than those shown in drawings 1 to 6.

2 Drawings 7 to 22 are symbols used by the firm Siemens Limited to identify electrical fittings and appliances.
Make a list stating which fitting or appliance you think each of the sixteen symbols represents.
Make accurate drawings of the sixteen symbols, working to any scale you find convenient.

3 Drawing 23 is an ideogram showing a telephone. You will find a variety of telephone ideograms used in advertisements of all kinds.
Draw a similar telephone ideogram of your own design. Make an accurate drawing of your design using instruments. Tint your drawing with coloured crayons or with water colour paint.

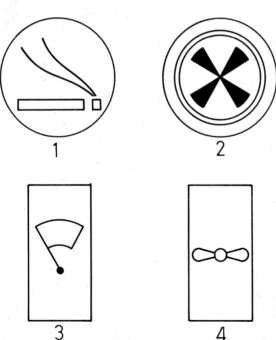

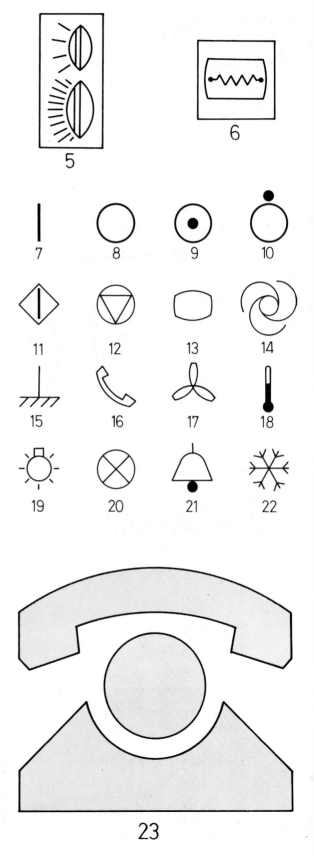

34

EITB Standard Symbols

Permission to reproduce the symbols shown on this page has been given by the Engineering Industry Training Board, who are the owners of the copyright vested in the symbols.

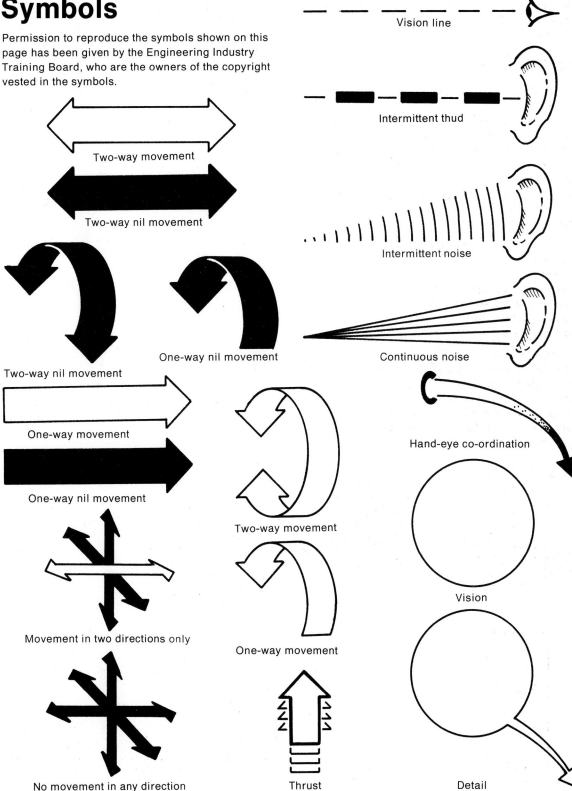

Leader line

Vision line

Intermittent thud

Intermittent noise

Continuous noise

Two-way movement

Two-way nil movement

Two-way nil movement

One-way nil movement

One-way movement

One-way nil movement

Movement in two directions only

No movement in any direction

Two-way movement

One-way movement

Thrust

Hand-eye co-ordination

Vision

Detail

Protractor Practice

The four star constellations shown on this page can be seen in the sky by observers in the Northern hemisphere. The Plough (part of the Great Bear) and Little Bear can be seen during clear evenings at all times of the year. Orion is best seen during evenings in February. Cygnus is best seen during evenings in November.

The Pole Star (Polaris) is always no more than 2° from True North.

The Pointers of the Plough are so called because they point to the Pole Star.

Exercise

Use A3 paper.

Make accurate scale 1:1 drawings of the four constellations shown using a protractor to plot the angles.

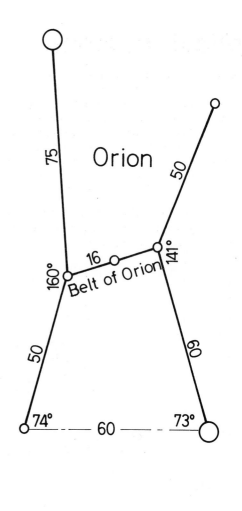

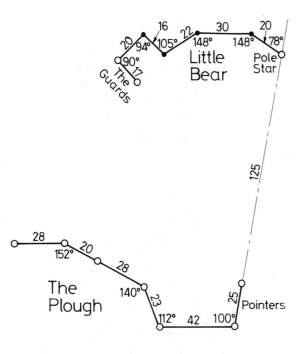

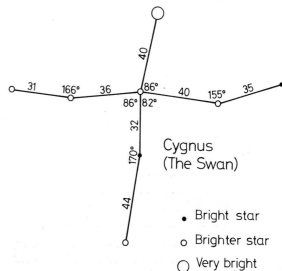

Optical Illusions

The seven drawings on this page are optical illusions. They are drawings which, on close examination, cause the eyes to see something different from what is actually drawn, or are meaningless, in that the articles portrayed cannot in fact be made. You may find it of interest to look for other optical illusion drawings. There are said to be about two hundred of them to be found in books of various kinds.

Drawing 1   The two vertical parallel lines separate parts of a straight sloping line. Do the two parts of the sloping line appear to be in line?

Drawing 2   The two vertical lines are parallel. Yet to most people they appear curved – closer at top and bottom, wider apart at the centre.

Drawing 3   The tips of the arrows divide the vertical line into two equal parts. Do the two parts appear to be equal?

Drawing 4   The upper 'scaled' portion of the vertical line is, in fact, exactly the same length as the lower unscaled part. Do the two parts appear to be equal?

Drawings 5, 6 and 7   Can these three articles be made? Yet they appear solid enough.

Exercise

Using instruments make accurate copies of the seven optical illusions using any scale you find suitable.

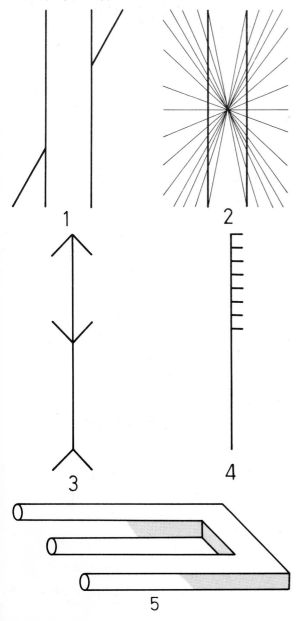

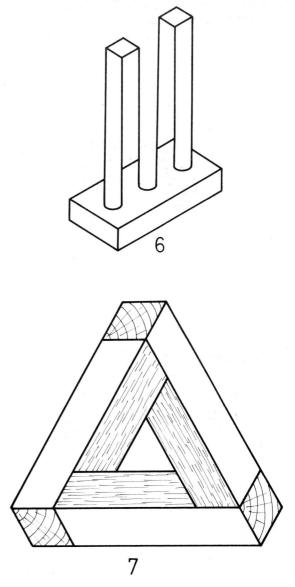

Graphs and Charts

Graphs and charts are forms of graphical communication for the conversion of tables of figures and other statistical details into drawings for easier and more rapid understanding. The types of graphs and charts illustrated in this pair of books are:

1 **Line graphs**   Included here to show the relationship between the graphical forms of parabolas and hyperbolas to the mathematical forms in graphs. See also page 21 of Book 2.

2 **Bar charts**   Vertical (histograms) and horizontal as well as block charts are shown (see p. 39).

3 **Pie charts**   Of circular form (see p. 40).

4 **Family trees**   See p. 41.

5 **Organization charts**   See p. 42.

6 **Flow charts**   See Book 2, pages 34 to 39.

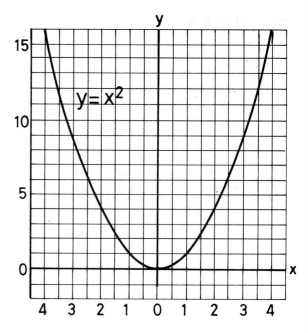

Line graphs

1 A line graph of the form $y = x$. This straight line graph shows the relationship between miles and kilometres. To convert miles into kilometres, follow the horizontal through the required mile position on the left-hand ordinate. Where this horizontal meets the graph line, read off the equivalent in kilometres vertically below.

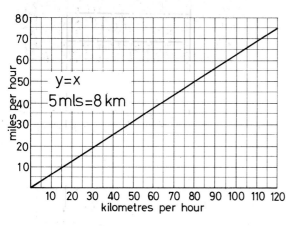

2 A line graph of the form $y = x^2$. The line is drawn through points found by substituting figures in the equation. Thus when $x = 1$, $y = 1$; when $x = 2$, $y = 4$; when $x = 3$, $y = 9$; when $x = 4$, $y = 16$. Also when $x = -1$, $y = 1$; when $x = -2$, $y = 4$; when $x = -3$, $y = 9$; when $x = -4$, $y = 16$. Draw a line through the points plotted. The graph of $y = x^2$ is a parabola.

3 A line graph of the form $y = \frac{1}{x}$. Plot points for the line are found by substituting in the equation. Thus when $x = 1$, $y = 1$; when $x = 2$, $y = \frac{1}{2}$; when $x = 3$, $y = \frac{1}{3}$; when $x = 4$, $y = \frac{1}{4}$. Also when $x = -1$, $y = -1$; when $x = -2$, $y = -\frac{1}{2}$; when $x = -3$, $y = -\frac{1}{3}$; when $x = -4$, $y = -\frac{1}{4}$. The resulting curve is a hyperbola. Note that the curve is in two separate parts.

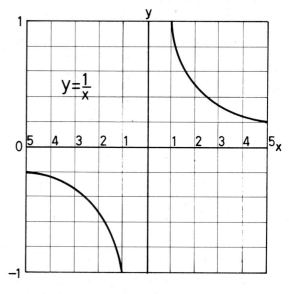

Exercises

Draw the three line graphs on graph paper. If graph paper is not available the graphs can be drawn on plain paper after constructing your own grid lines.

Bar charts

Six bar charts are shown on this page and on page 40. The student is advised to draw these to the details given below:

1 A horizontal bar chart showing a comparison between the melting points of twelve metals. The melting points, to the nearest 10° Centigrade, were taken as: Tin 230°; Bismuth 270°; Lead 330°; Zinc 420°; Aluminium 660°; Silver 960°; Gold 1060°; Copper 1080°; Nickel 1450°; Iron 1530°; Platinum 1770°; Tungsten 3370°.
The bars are 10 mm wide and the scale chosen for the base line is 5 mm represents 100°C.

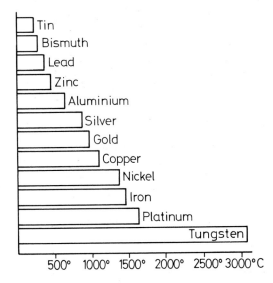

2 A vertical bar chart (a histogram) showing the increases in total numbers of candidates entered for C.S.E. examinations during the years 1970 to 1975.
The numbers of candidates are only approximate,

being measured in thousands, so the numbers can be estimated from the given chart. Use the following dimensions:

Bars 10 mm wide; 10 mm represents each group of 100,000 candidates. Tint the girl columns with pencil crayon or with water paint.

3 A vertical bar chart showing the value of stores of materials held by a comprehensive school design and craft department at the end of each month during the past year. The height of each component of the bars can be assessed from the given drawing. Otherwise work to the following dimensions:

Each bar 8 mm wide; each £1000 represented by 20 mm height; tint and colour with crayon or paints. Answer the following questions relating to this chart:

(a) Why are the July and August stocks the same?
(b) Why is a sudden large decrease in stocks shown at the ends of June and November?
(c) Account for the sudden large increases in stocks in July and December.

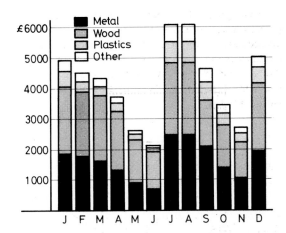

4 This histogram shows the losses and profits for an industrial firm for the years 1978 to 1984. Copy the chart to suitable dimensions of your own choice.

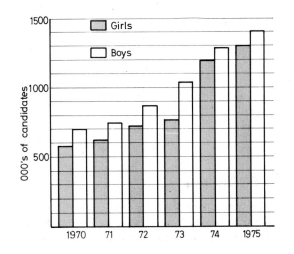

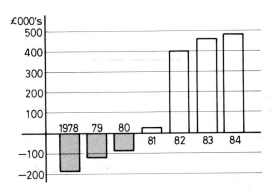

5   A horizontal bar chart showing comparisons between the braking distances of a motor car driven at speeds from 40 km/h to 100 km/h (kilometres per hour). The distances were taken from average reaction times, braking on level, dry roads.
Copy the diagram using your own dimensions.

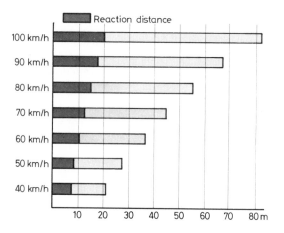

6   A block chart showing comparisons between the various constituents of food taken daily by a middle-class American.
Copy and tint the diagram. Make the block 60 mm wide and 100 mm high overall.

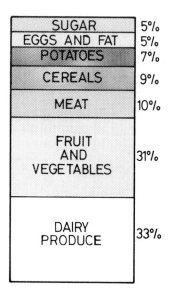

Pie charts

Two pie charts are shown. Pie charts are circular with radii drawn between the various parts. To draw a pie chart it is necessary to work out the angles of each of the sectors. The first of the two pie charts shown repeats the information given in block chart 6, but in a different form. Thus 33% of the circle showing the consumption of dairy produce daily requires a sector of 33% × 360° = 119°; the fruit and vegetable sector is 31% × 360° = 112°; the meat sector is 10% × 360° = 36°; and so on.
Copy the pie chart and tint your drawing.
The second of the two pie charts compares the values of imports into the various areas of the British Isles during a certain year. The figures, rounded up to the nearest £1 000 000, are as follows.

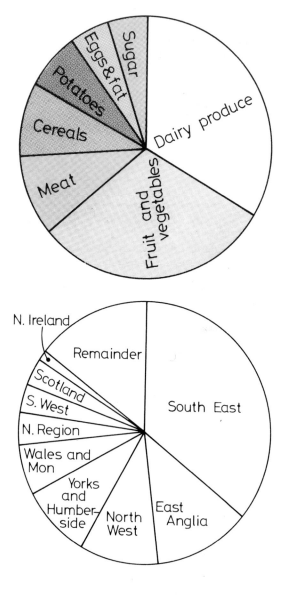

The angles of the sectors should be worked out by the student.

1.	South East	£3 195 000 000	approx. 36%
2.	East Anglia	£1 102 000 000	12%
3.	North West	£ 907 000 000	10%
4.	Yorks and Humberside	£ 806 000 000	9%
5.	Wales and Monmouthshire	£ 538 000 000	6%
6.	North Region	£ 350 000 000	4%
7.	South West	£ 325 000 000	4%
8.	Scotland	£ 278 000 000	3%
9.	Northern Ireland	£ 99 000 000	1%
10.	Remainder	£1 335 000 000	15%

Family trees

The family tree drawn on this page illustrates the relationship between the eight cousins Francis, Alison, William, Victor, Marion, John, Kathy and Claire, with their four grandparents Albert and Margaret Smith and Edward and Jane Wood.

Exercise

1 Copy the given family tree.

2 Design and draw a family tree showing your own relationship with your cousins and your grandparents as far as you are able to trace these relationships.

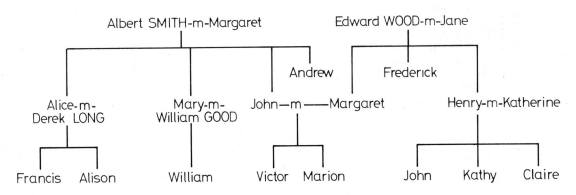

Organization charts

The large diagram on this page shows how a school is organized from the Department of Education and Science. The school illustrated in this chart is a large comprehensive school of some 2000 pupils whose ages range from 11 years up to 19 years. Part of the chart is shaded. This shading shows the Design faculty. It is within the Design faculty of this particular school that Graphical Communication is taught. Under each of the departments within the Design faculty a variety of subjects will be taught.

Copy the shaded part of the chart to a larger scale and add to your drawing the organization of the subjects within each of the departments – Technical, Art, Domestic and Commerce.

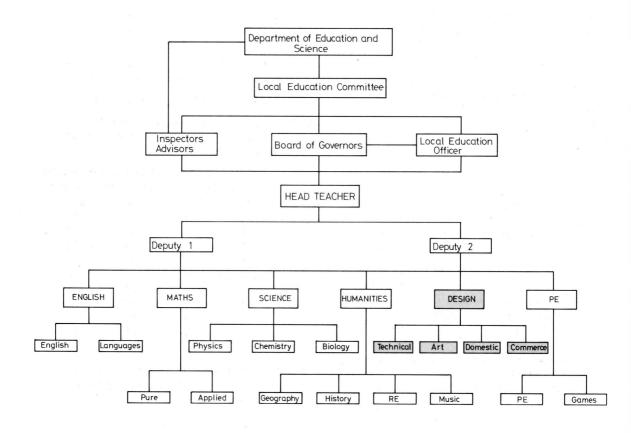

Exercises

1 The following percentages are the proportions of net incomes spent on each item by the average family.

Item	% of Net Income
Food	30
Housing	20
Fuel	12·5
Transport	15
Clothes	7·5
Furniture	6
Other things	9

(a) Construct a pie chart showing each item as a percentage of the net income.

(b) If a man has a net income of £5600 per annum, using the above information draw a bar chart to show how much he would spend on each item.

(A.E.B.)

2 A shop is holding a sale and the manager wishes to ensure that all out-dated stock is sold. A stock check has been carried out on the values of the items to be cleared. This is listed below.

Item	Value in £	% of Total Value
Wood panels	600	
Ceiling tiles	900	
Floor tiles	750	
Shelf units	450	
Dustbins	300	

Make a list showing the percentages of value for each item and construct a pie chart showing the value of each item as a percentage of the total value of the out-dated stock. Colour should be used.

(A.E.B.)

3 Albert Young married Anne Smith at the end of the nineteenth century. Their five children in the order of their birth were: Albert (died aged 3 years); George; Edward (died aged 6 months); Anne and Elizabeth. The three surviving children married as follows:

George married Winifred Smart – children Louise, Thomas and Alfred.

Anne married Bryan Goodall – children Joyce and Richard.

Elizabeth married Richard Bath – no children. Louise remained a spinster; Thomas married his cousin Joyce Goodall; Alfred married Rosalind Howard; Richard married Joan Bryant.

There are now four great-grandchildren of Albert and Anne Young. Stephen and Anne Goodall are at a secondary school; John and Joan Young are in a primary school.

Construct a family 'tree' showing the relationships between all the persons named above.

Colour may be used as appropriate.

4 Construct a bar chart showing comparisons between the densities of the following metals:

Metal	Density – grams per cubic cm
Aluminium	2·7
Bismuth	9·8
Copper	8·9
Gold	19·3
Iron	7·9
Lead	11·3
Nickel	8·9
Platinum	21.5
Silver	10·5
Tungsten	19·3
Zinc	7·1

Colour may be used if thought appropriate.

5 In answer to a questionnaire from a local education authority ten secondary schools supplied the following information:

School	5 subjects or more		Less than 5 subjects	
	Boys	Girls	Boys	Girls
1	56	53	15	28
2	60	57	29	18
3	72	45	18	56
4	86	80	34	41
5	89	91	20	40
6	105	101	28	56
7	107	109	16	54
8	110	103	60	64
9	120	114	68	60
10	115	124	54	80

(Column header above the table: *Pupils in 5th Year entered for examinations*)

Design and draw a bar chart which will convey the above information in a graphical form. Colour should be used.

Golden Mean Proportions

Divide a line into 'golden mean' proportions.
Line AB is 90 mm long.

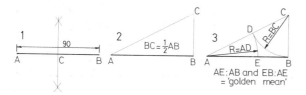

Golden mean rectangle

Draw a golden mean rectangle.
AB is 90 mm long.

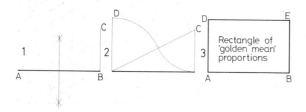

Application to layout

A golden mean rectangle within which a circle and an ellipse are drawn at positions relating to golden mean proportions.

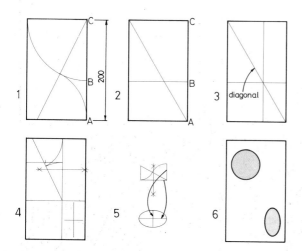

Geometrical Patterns

The design of geometrical patterns is an interesting exercise in graphical communication. Such patterns are formed from lines drawn with the aid of instruments – Tee square, set squares, compasses and curve aids. The addition of colour by shading or tinting or the colouring of lines can enhance the appearance of the designs if added with care and discretion.

Six geometrical patterns are shown on page 45 opposite. These have all been drawn within squares of 85 mm side lengths. All six can be drawn on a single sheet of paper size A4. The pieces of isometric grid paper for patterns 3 and 5 can be attached to the drawing sheet with paste, glue or self-adhesive tape.

1   A line pattern consisting of a number of touching squares. Alternating squares are at 45° to each other.

2   A pattern forming a cross. Drawn with the aid of a 45° set square.

3   This pattern is drawn on 5 mm isometric grid paper. Such designs can be seen in the arrangements of blocks in parquet flooring.

4   Vertical lines and lines at 60° drawn at intervals of 20 mm form the basis for this design. Alternate triangles may be shaded with a clear colour wash.

5   Another pattern drawn on an isometric grid. This time the grid is spaced at 10 mm spacing intervals.

6   A line pattern. Three points have been taken, more or less at random, outside the 85 mm square. Lines drawn from these three points intersect within the square. Some of the lines could be coloured.

Exercise

Draw the six patterns given on page 45. Also design patterns based on those given.

85 mm
squares

20

5 mm Isometric
grid

1

2

3

20

4

5

10 mm Isometric
grid

6

Geometrical patterns

Twelve more geometrical patterns drawn within squares are shown on page 47 opposite. The grids on which these twelve patterns are based are given on this page.

Exercises

Copy the twelve patterns on page 46. Commence by drawing the grids using fine construction lines drawn with the aid of a 45° set square. Add colour if thought appropriate.

Design other patterns based on the ideas contained in these twelve designs. Some of the designs shown were taken from the book *A Treasury of Design for Artists* by Gregory Mirow, published by Dover Publications Incorporated of New York. If you can obtain a copy of this book you will find a very large number of interesting geometrical designs within its covers.

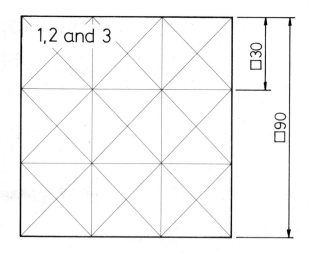

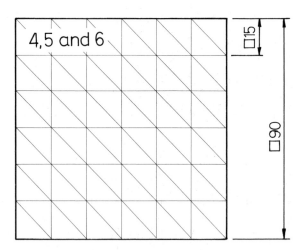

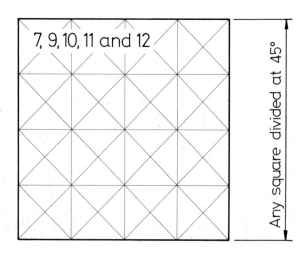

Thread Patterns

The line patterns shown here are based on the design work produced by interlacing coloured threads between nails on wooden boards. In the case of the actual thread patterns, the threads are attached to the boards by nails which are spaced at regular intervals along lines drawn on the backing boards.

The thread patterns on this page are based on lines drawn between regularly spaced points measured along lines drawn on the paper.

The patterns can be drawn as exercises and can be coloured if this is thought to be appropriate. When the six patterns have been copied, the student is advised to attempt new designs for himself.

4

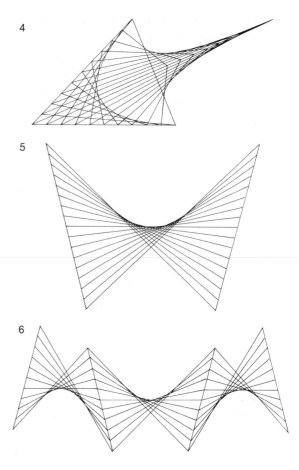

5

6

1

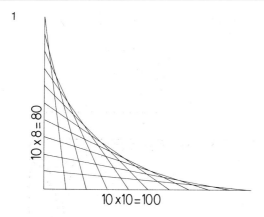

$10 \times 8 = 80$

$10 \times 10 = 100$

2

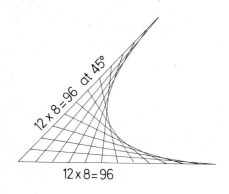

$12 \times 8 = 96$ at $45°$

$12 \times 8 = 96$

3

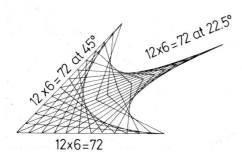

$12 \times 6 = 72$ at $45°$

$12 \times 6 = 72$ at $22.5°$

$12 \times 6 = 72$ at $45°$

$12 \times 6 = 72$

Exercises

1　Two lines at 90°. Spaces along base line at 10 mm intervals. Spaces along vertical at 8 mm intervals.

2　Lines at 45°. Spacing along both lines at 8 mm intervals.

3　Two lines at 45°. A third line along the bisector of the 45° angle. All spaces at 6 mm intervals.

4　Similar to 3 with bisector line spacing in a different position along the bisector.

5　Two lines at opposite angles. Choose any spacing.

6　Similar to 5 with four sloping lines.

7　The four lines on which the thread pattern shown cn the colour page opposite is drawn.

7

Each $12 \times 6 = 72$

48

A thread pattern on the lines spaced as shown in drawing 7 on page 48 opposite (*upper*).

A geometrical pattern design composed of lines, circles and ellipses. Different shades of colour have been used on this design (*lower*).

Exercises

1 Draw the thread patterns shown.

2 Copy the geometrical pattern. The borders of the pattern are within a rectangle measuring 240 mm by 190 mm.

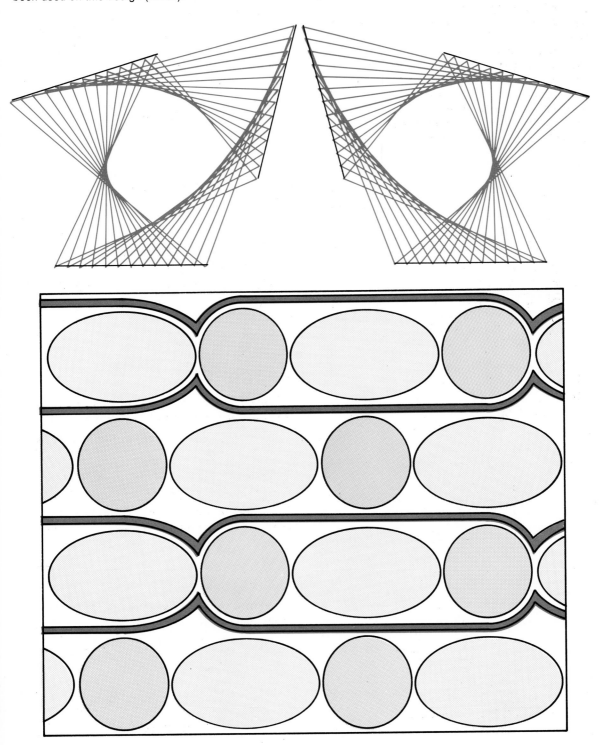

Garage extension drawing

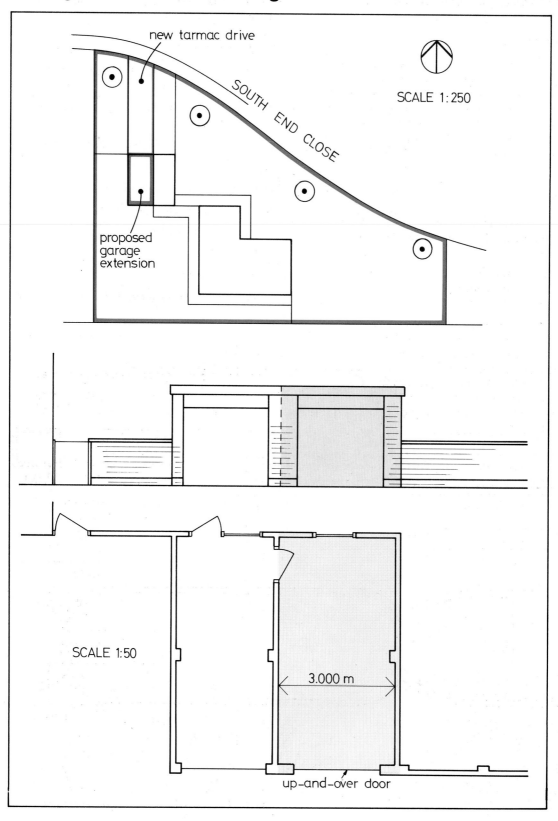

new tarmac drive

SOUTH END CLOSE

SCALE 1:250

proposed garage extension

SCALE 1:50

3.000 m

up-and-over door

Building Drawing

Symbols

Many symbols and abbreviations are used in the making of building drawings. Students are advised to refer to British Standard 1192 (Recommendations for Building Drawing Practice) for complete lists of these. The symbols and abbreviations employed in drawings in this book and in Book 2 are shown here. This list of symbols is not shown in alphabetical order.

Lines

Three thicknesses of lines should be used in building drawings. British Standard 1192 recommends line thicknesses of:

0·8 mm wide – Thick lines
0·4 mm wide – Medium lines
0·2 mm wide – Thin lines

When working with pencil as in most school drawings, such line thicknesses are difficult to achieve. However, in pencil drawings, thick lines should be approximately twice the width of medium lines and medium lines should be approximately twice the width of thin lines.

Precise line thicknesses for building drawings can be obtained when using technical pens of the Rotring type (see Book 2). Lines should appear on drawings as follows:

Site location drawings

Thick lines — outlines of plot; outlines of buildings on the site

Medium lines — outlines of buildings in the vicinity of the site

Thin lines — plot boundaries, road boundaries

Site plans

Thick lines — outlines of buildings

Medium lines — boundaries, drain lines

Thin lines — dimensions

Building plans

Thick lines — walls

Medium lines — windows, doors, fixtures

Thin lines — grids

Scales

Recommended scales for building drawings are:

1:2500	1:1250	1:500	1:200	1:100
1:50	1:20	1:10	1:5	1:1

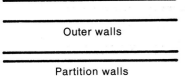
Outer walls

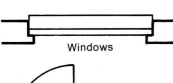
Partition walls

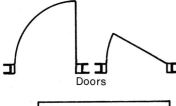
Windows

Doors

Radiators (central heating)

True North

MH
Manholes (Soil)

MH
Manholes (Surface water)

RWP
Rainwater pipes

Baths

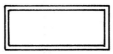
S
Sinks

WB
Wash basins

B
Boilers

C
Cookers

R
Refrigerators

Water closets

Types of drawing

Many different types of drawings are used in connection with building. Only three of these types will be considered in this book and in Book 2. These are:

1 Site location plans (see below, page 51)

2 Site plans (see below, page 52)

3 Building plans (see below, page 53)

Other types of building drawings are beyond the scope of this book. Thus the following types of drawings are not included:

Surveyor's plots; drawings of building detail; component drawings; assembly drawings; installation drawings; block plans; schedule drawings.

Site location plan

On page 51 opposite are two drawings, both drawn on A4 size drawing sheets. The upper drawing was made freehand at the site on which the building known as 2 Burton Road was eventually built. Such freehand drawings would normally be made from surveyor's plots, but we are not concerned with surveying in this book.
The lower drawing is a scale 1:1250 site location plan for the plot on which 2 Burton Road was built. This drawing was made with the aid of instruments.

Note Even when making building drawings freehand, line thickness variations should be observed. Thus the plot outlines in the freehand drawing are drawn with thick lines, medium lines show the outlines of other buildings in the vicinity, while thin lines show field and plot boundaries of other buildings.

Exercises

1 On sheets of A4 paper make copies of the two drawings on page 51.

2 Make a pair of similar drawings of site location plans for the house where you are presently living.

Site plan

On page 52 a site plan is drawn for the same building as that shown in the site location plan on page 51. This site plan has been drawn on an A3 sheet to a scale of 1:200. It includes only those parts of the area immediately associated with the building to be known as 2 Burton Road. Note the three thicknesses of lines,

following the rules laid down on page 49. Note also the method of showing dimensions on building drawings of this nature. Dimension lines are thin and terminate in arrows drawn with the aid of a 45° set square. Dimensions are taken to three decimal places to avoid any possible confusion with millimetre dimensions which may be placed on the same drawing. Also note the true North symbol. To make it easier to draw the building outlines, the paper has been set up parallel to the drawing board edges. This has meant that the North symbol has had to be rotated to conform with the compass position in relation to the building.

Exercises

1 Copy the drawing on page 52 on a sheet of A3 paper.

2 Make a similar site plan from the site location plans for your own dwelling.

Building plan

A building plan for 2 Burton Road is given on page 53. Note the following:

1 Thickness of lines conforms to the rules laid down on page 49.

2 Rooms are numbered clockwise around the plan.

3 Windows and doors are numbered clockwise around the plan.

4 Dimensions of rooms taken from a freehand drawing made on grid paper.

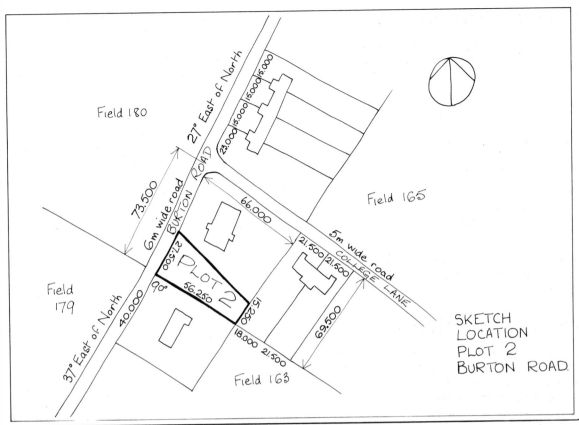

Field 180

27° East of North

BURTON ROAD

6m wide road

73.500

23.000|5.000|5.000|5.000

Field 165

66.000

21.500|21.500

5m wide road

COLLEGE LANE

Field 179

37° East of North

40.000

90°

27.500

PLOT 2

56.250

15.250

18.000   21.500

69.500

Field 163

SKETCH
LOCATION
PLOT 2
BURTON ROAD.

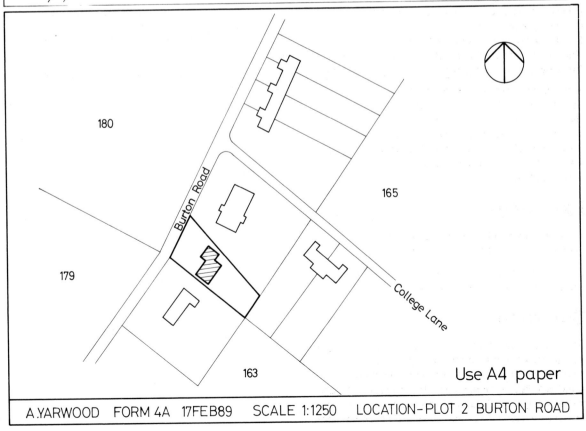

180

Burton Road

165

179

College Lane

163

Use A4 paper

A. YARWOOD   FORM 4A   17 FEB 89   SCALE 1:1250   LOCATION-PLOT 2 BURTON ROAD

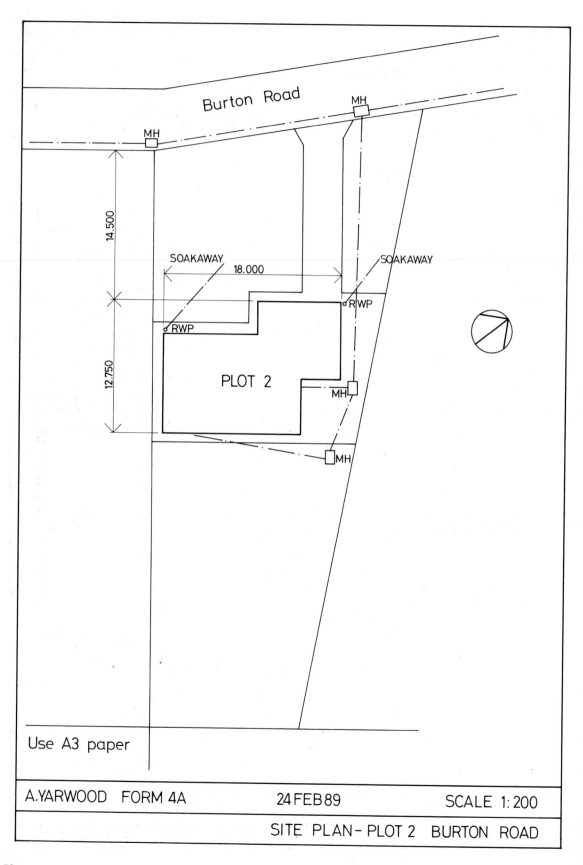

Burton Road

MH

MH

14.500

SOAKAWAY

18.000

SOAKAWAY

RWP

RWP

PLOT 2

MH

MH

12.750

Use A3 paper

A.YARWOOD   FORM 4A   24 FEB 89   SCALE 1:200

SITE PLAN – PLOT 2   BURTON ROAD

Building plan

A scaled, freehand drawing made on a 1 metre grid is shown. Copy this drawing on square grid paper and also make a similar drawing for the house in which you live.

The photograph shows a scale model of 2 Burton Road made from different coloured cards. Scale model making is a useful exercise to practice.

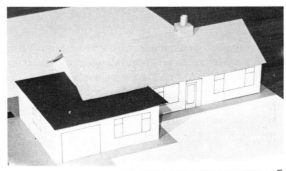

Dimensions for 2 Burton Road

O1	Living room	8·900 m × 5·400 m
O2	Bedroom or study	3·350 m × 2·400 m
O3	Bedroom	5·650 m × 3·750 m
O4	Garage	6·500 m × 3·400 m
O5	Bathroom	2·350 m × 2·000 m
O6	Bedroom	4·000 m × 3·750 m
O7	Kitchen	4·150 m × 4·000 m
O8	Hall	
W5		600 mm wide
W2, W4, W6, W7		1200 mm wide
W1, W3, W8		1800 mm wide
All doors except D2		1·000 m wide
D2		2·100 m wide

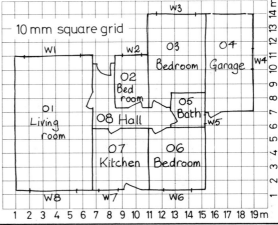

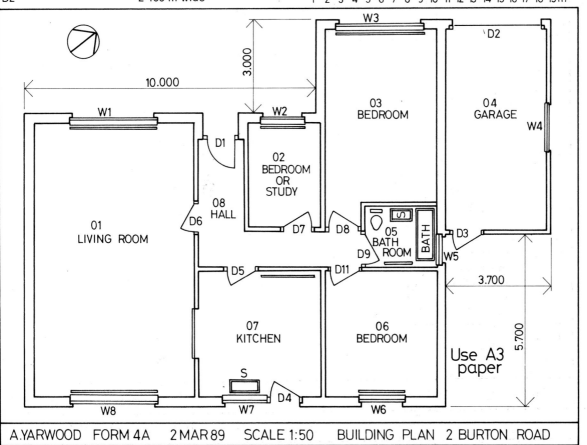

A. YARWOOD   FORM 4A   2 MAR 89   SCALE 1:50   BUILDING PLAN 2 BURTON ROAD

Vectors

A quantity having magnitude but no direction is called a **scalar** quantity. A quantity which has both magnitude and direction is called a **vector** quantity. Vectors may be shown graphically by lines. The scaled length of the line gives magnitude. The position of the line gives direction.

Drawing 1   A displacement from A to B of 8 metres can be shown by the vector AB.

Drawing 2   C is a vector representing a south-westerly wind of 15 km per hour.

Drawing 3   D is a vector illustrating a force of 4 newtons acting vertically downwards.

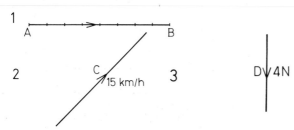

Example 1

A man paces 6 metres in a north-westerly direction. He then turns east and paces 5 metres. Find the actual distance the man has moved and his direction from his starting point.

Answer 1   To a scale of 1:100 draw ab (vector A) and bc (vector B) parallel to the directions in which the man has moved. Then ac (vector C) indicates the man's distance to and his present direction from the starting point. In triangle abc vector A + vector B = vector C. The addition of vectors is not arithmetical. By measurement, vector C (ac) = 43 mm. Therefore the distance moved is 4·3 m at 10° E of N from the start.

Note   1   abc is a complete triangle.
2   The arrows of vectors A and B are clockwise around the triangle. The arrow of vector C is anti-clockwise. Vector C is in the opposite sense to vectors A and B.

Example 1

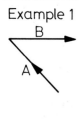

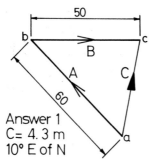

Answer 1
C= 4.3 m
10° E of N

Example 2

A dinghy sails in a north-easterly direction for a distance of 4 km. The wind changes and the dinghy then sails a further 6 km in a west-north-west direction. How far is the dinghy from its starting point? What is the bearing of the dinghy from the start?

Answer 2   To a scale of 10 mm represents 1 km draw vector triangle abc. Find A + B = C by measuring vector ac. Answer: 5·9 km on a bearing 26° west of north.

Example 2

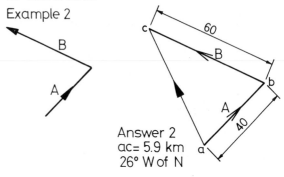

Answer 2
ac= 5.9 km
26° W of N

Example 3

A 10 newton force acts at a point at right angles to a 15 newton force. What single force could replace the two forces?

Answer 3   Working to a scale of 5 mm represents 1 newton draw the vector diagram ab and ad. Draw bc parallel to vector A and cd parallel to vector B to obtain the parallelogram of forces abcd. Note that this parallelogram can be regarded as consisting of two identical triangles abc and acd. In either of these triangles of force the vector R (ac) represents the single force which could replace the 10 N and 15 N forces. R represents the **resultant** of the two forces. Also A + B = R in vector addition.

Example 3

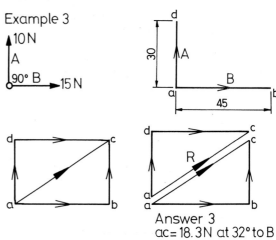

Answer 3
ac= 18.3N at 32° to B

Example 4

Find the resultant of the two forces.

Answer 4 Scale 2 mm represents 1 newton. Draw ab parallel to the 30 N force and bc parallel to the 20 N force. Vector R gives the required resultant. On measurement this is 21 N at 43° to the 30 N force.

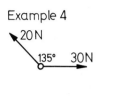

Example 4

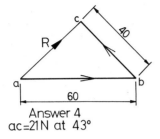

Answer 4
ac=21N at 43°

Example 5

Find the resultant of the two forces.

Answer 5 By measurement from vector diagram abc.

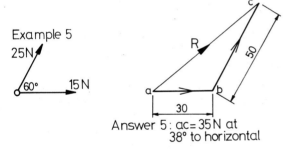

Answer 5: ac=35N at
38° to horizontal

Polygon of forces

Straight line polygons consist of adjoining triangles. The given pentagon can be subdivided into three triangles. Because of this, the system of triangles of vectors can be extended to the resolution of forces which form polygons of vectors.

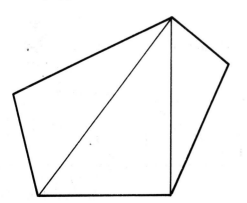

Example 1

Find the resultant of the three forces.

Answer 1 Scale 100 mm represents 1 kilonewton. Vector ab is parallel to the 0·4 kN force, vector bc parallel to the 0·3 kN force and vector cd parallel to the 0·6 kN force. The closing vector ad is the vector representing the resultant of the three forces.

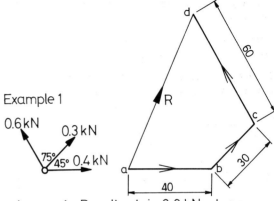

Answer 1: Resultant is 0.8 kN at an
angle 66° to the 0.4 kN
force

Example 2

In which direction will the system of forces be tending to move? Find the resultant of the four forces.

Answer 2 By measurement of vector ae.

Note In the two polygons of force shown:

1 The resultant is opposite in sense to the given forces.

2 Both polygons are closed by the resultants.

Example 2

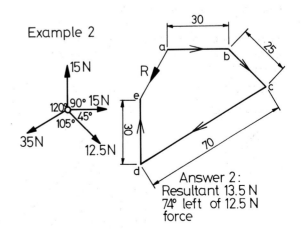

Answer 2:
Resultant 13.5 N
74° left of 12.5 N
force

Force Diagrams

The worked examples on this page show how force diagrams can be laid out on a sheet of A4 paper. The examples given on pages 54 and 55 can also be laid out on A4 paper in a similar manner.

Example 1

Find the magnitude of the forces X and Y which will hold the point stationary.
Draw vector ab parallel to the 10 N force. Draw vector ca parallel to the unknown force Y. Draw bc parallel to the unknown force X. Measurement of vector ca and bc gives the magnitude of forces X and Y = 7·4 N and 5·2 N.

Example 2

Find the magnitude of the forces X and Y which will hold the point stationary.
Draw vector ab parallel to the 10 N force. Draw vectors bc and ca parallel to the unknown forces X and Y. Measure the scaled lengths bc and ca.

Example 3

Find the resultant of the 7 N and 9 N forces.

Draw vector ab 70 mm long parallel to the 7 N force.
Draw vector bc 90 mm long parallel to the 9 N force.
Measure vector ac = 11·5 N at 7° to horizontal.

Example 4

Find forces X and Y which will hold the point stationary.
Draw vector ab 100 mm long parallel to 10 N force.
Draw vector bc 105 mm long parallel to 10·5 N force.
From c draw a line parallel to force X. From a draw a line parallel to the force Y. By measurement of the two resulting vectors X = 5·2 N and Y = 2·9 N.
The point at which the four forces are acting is said to be 'in equilibrium' under the action of the four forces.

Note In diagrams 1, 2 and 4, the arrows showing the direction in which the forces are acting are all in the same sense around the triangles or polygon. If the forces are acting at a point, that point will be held stationary under the action of such forces.
In diagram 3, the arrow showing the direction of the resultant is in the opposite sense to the other two forces. The resultant in this case can replace the two given forces.

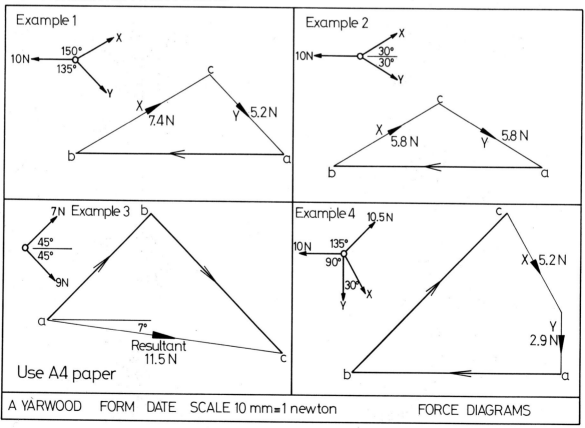

A YARWOOD   FORM   DATE   SCALE 10 mm ≡ 1 newton   FORCE DIAGRAMS

56

Exercises

1 By drawing a triangle of forces, determine the magnitude of the forces A and B which will hold P stationary.

2 Determine the magnitude of forces A and B necessary to hold P in equilibrium.

3 Draw a triangle of forces from which the magnitude of forces A and B necessary to hold P in equilibrium can be found.

4 Find the resultant of the two forces.

5 Find the resultant of the 4 N and 7 N forces.

6 Find the resultant of the 75 N and 100 N forces.

7 Determine by drawing a polygon of forces the forces A and B which will hold the given two forces in equilibrium.

8 Find the forces A and B necessary to hold P stationary.

9 Find the forces A and B which are needed to hold P stationary under the action of the 5 N and 90 N forces.

10 Draw a pentagon of forces to find the magnitude of forces A and B needed to hold P in equilibrium under the action of all five forces.

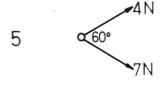

5

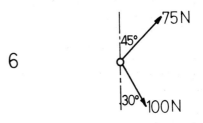

6

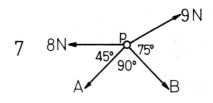

7

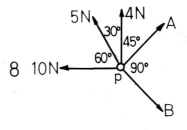

8

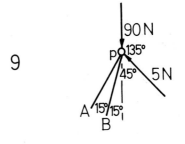

9

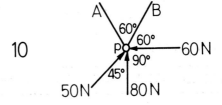

10

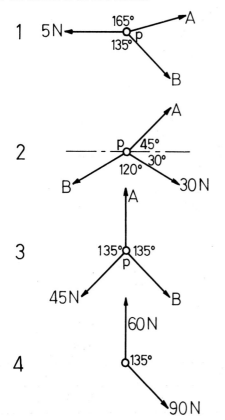

Electrical Circuits

British Standard 3939 (Graphical symbols for electrical power, telecommunications and electronics diagrams) recommends symbols and conventions for the drawing of electrical circuits. BS 3939 is a complex standard because of the large number of symbols required in the increasingly complicated circuits in present-day electrical installations. The symbols necessary for the student to be able to draw simple electrical and electronics circuits are shown on this page. These are based on the recommendations of BS 3939. For the symbols used in more complex circuits the student is advised to consult BS 3939.

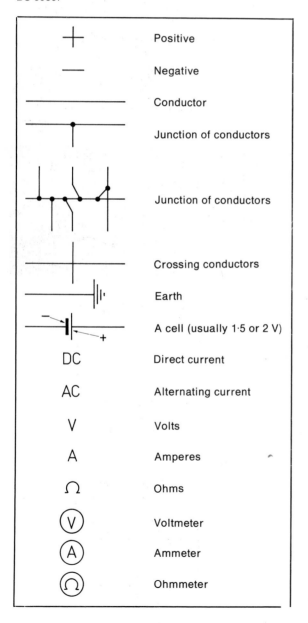

	Positive
	Negative
	Conductor
	Junction of conductors
	Junction of conductors
	Crossing conductors
	Earth
	A cell (usually 1·5 or 2 V)
DC	Direct current
AC	Alternating current
V	Volts
A	Amperes
Ω	Ohms
V	Voltmeter
A	Ammeter
Ω	Ohmmeter

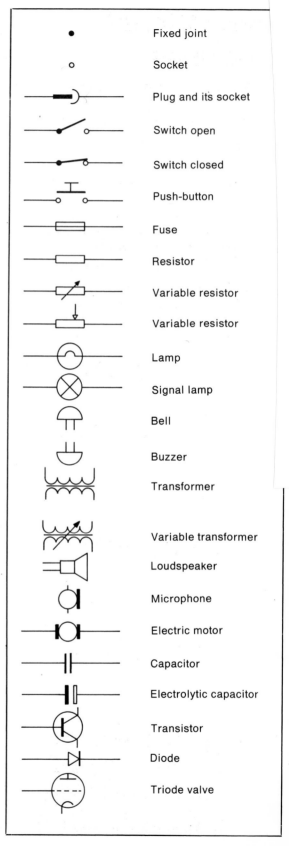

	Fixed joint
	Socket
	Plug and its socket
	Switch open
	Switch closed
	Push-button
	Fuse
	Resistor
	Variable resistor
	Variable resistor
	Lamp
	Signal lamp
	Bell
	Buzzer
	Transformer
	Variable transformer
	Loudspeaker
	Microphone
	Electric motor
	Capacitor
	Electrolytic capacitor
	Transistor
	Diode
	Triode valve

Some Circuit Diagrams

Six elementary circuit diagrams are shown. These are:

1   A 6 V battery connected to a switch and lamp.

2   A 6 V battery connected to a switch and to two lamps in series.

3   A 6 V battery connected to a switch and to two lamps in parallel.

4   The electrical circuit diagram for a polystyrene foam hot wire cutter. A sketch of the components as they are fitted in the cutter is also shown.

5   Part of a school experiment to test current flow in and voltage drop across a resistance wire.

6   Part of a school experiment showing how a school-made bi-metallic strip may be used as a two-way switch.

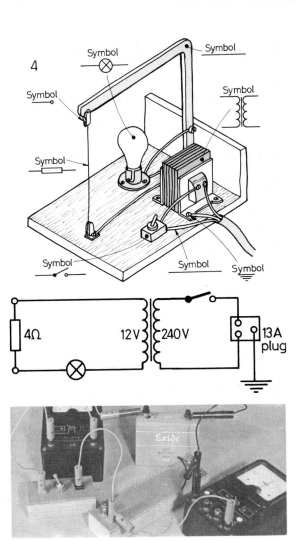

4

Symbol

Symbol

Symbol

Symbol

Symbol

Symbol

Symbol

Symbol

4Ω   12V ⫸ ⫷240V   13A plug

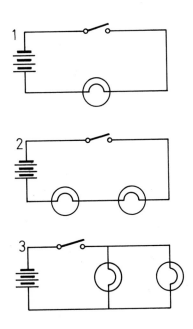

1

2

3

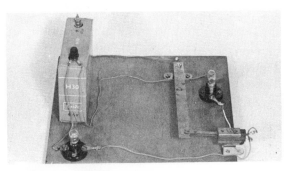

5   A

V

6

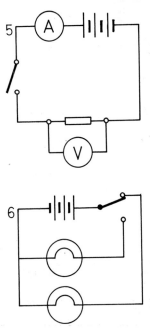

Motor car circuits

Electrical circuits may be drawn freehand on square grid paper in preparation for more exact drawings with instruments. Drawing 1 is an example of such a freehand drawing and drawing 3 shows the finished drawing. Freehand sketches such as drawing 2 may also be prepared as a preliminary to a finished circuit diagram. In drawing 2 the symbols have been replaced with the names of the components.
Drawing 4 is a circuit diagram for the wiring of a horn in a motor car. Note the completion of the circuit through 'earth' connections. The metal of the car body is used as an 'earth' to complete the electrical circuit.

Exercise

Copy the diagrams given on this page and on page 59. Use any convenient scale.

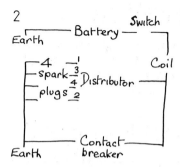

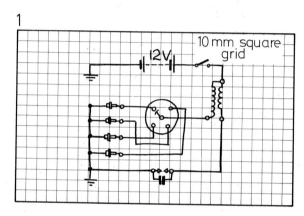

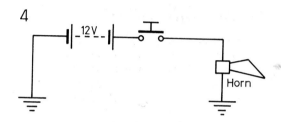

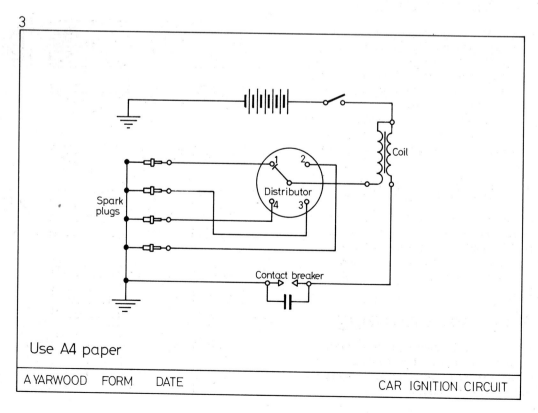

Use A4 paper

A YARWOOD   FORM   DATE   CAR IGNITION CIRCUIT

Methods of Graphical Illustration

Five methods of graphical illustration are described in this book and in Book 2. These methods are suitable for either instrument drawings or freehand drawings. The drawings against grey background are freehand.

1 Isometric drawing

A good, general method for the production of pictorial drawings (see page 62). Suitable for illustrating small components of all types. Can also be employed for larger articles if desired. 'Exploded' drawings are often isometric.

2 Cabinet drawing

A form of oblique drawing (see page 67). Tends to show distortions in some pictorial drawings. Very easy to construct. Suitable for drawings which contain curves in one view. After drawing a front view, lines are taken at half the drawing scale along 45° axes.

3 Planometric drawing

Suitable for drawings such as room layouts, building layouts, area planning drawings (see page 68). Vertical lines are taken from a plan to produce a three-dimensional view. Two methods are common – one drawn with the plan tilted through 45°, the second with the plan tilted through 30°. When the plan is tilted through 45° a shortening of verticals to $\frac{3}{4}$ (or $\frac{2}{3}$) of the drawing scale will result in a drawing showing less distortion than if full-scale verticals are drawn.

4 Orthographic projection

The most common form of graphical illustration (see page 70). Suitable for engineering and building drawings. Two 'angles' of projection – First Angle (the upper drawing) and Third Angle (the lower). Pages 70 onwards are devoted to orthographic projections.

5 Perspective drawing

Described in Book 2. Suitable for drawing room layouts, buildings or groups of buildings and other large objects. Two types are described, single point (parallel) and two point (angular).

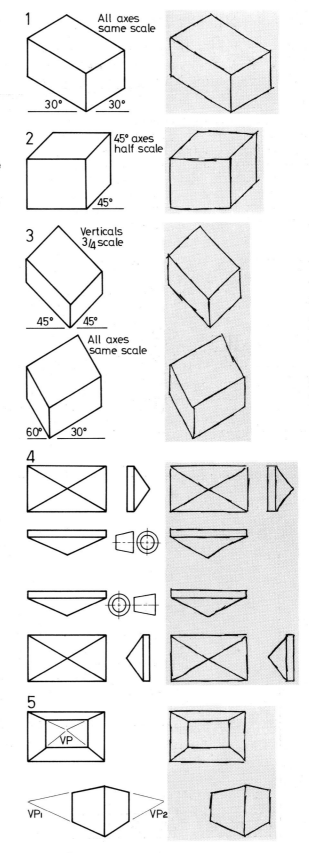

1 — All axes same scale — 30° 30°

2 — 45° axes half scale — 45°

3 — Verticals 3/4 scale — 45° 45°

All axes same scale — 60° 30°

4

5 — VP — VP₁ VP₂

Isometric Drawing

All lines in isometric drawings are drawn with the aid of a 30°, 60° set square. Determine the scale to which the drawing is to be made. Measurements to this chosen scale should only be made along vertical or 30° axes.

Commence an isometric drawing by constructing isometric 'boxes' – outlines drawn with thin fine lines – to the main dimensions of each part. Details of the drawing are then constructed within the 'box' outlines before lining in the parts needed for the finished isometric illustration.

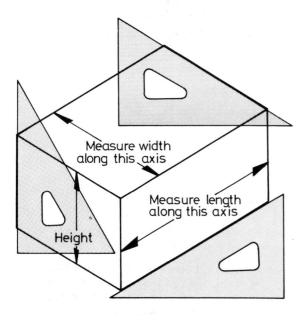

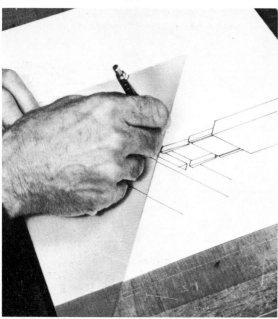

Exercises

1   Use an A4 sheet in 'portrait' position – with the short edges of the sheet horizontal. Make an isometric drawing of the bridle joint to a scale of 1:1. Add a border and title block to your drawing. Print your name, the scale used and TEE BRIDLE JOINT in the title block.

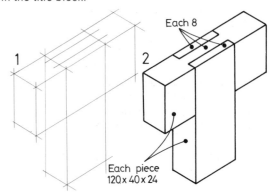

2   Use an A4 sheet in 'landscape' position – with the long edges of the sheet horizontal. Make an isometric drawing of the table frame to a scale of 1:5. Choose suitable dimensions for the rails and legs of the frame. Add a border and a title block to your drawing. Include your name, the scale and TABLE FRAME in the title block.

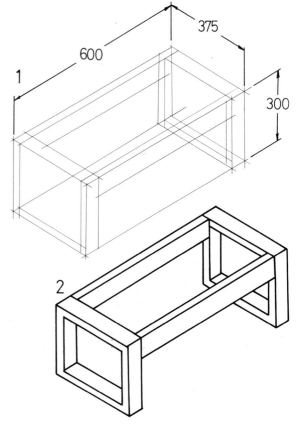

Curves in isometric drawings

To draw curves in an isometric drawing first draw the outline of the curve as seen looking directly at its shape. On this outline draw parallel straight line ordinates at intervals to cross the curve. Points on the curve in the isometric drawing are then plotted by repeating the ordinates on the isometric axes. The points where the curve crosses each ordinate are then plotted on the isometric ordinates. When a sufficient number of points have been plotted the curve is drawn freehand through the points.

This method is shown in the six drawings illustrating the making of an isometric drawing of a cylinder. The stages as shown in these six drawings are:

1   Draw ordinates across the circle of the cylinder.

2   Draw the isometric 'box' in which the isometric drawing of the cylinder is to be constructed.

3   Measure the distances 1 and 2 along the centre lines of the top and bottom of the cylinder. These should be the same as the distances 1 and 2 on drawing 1.

4   Draw the ordinates along 30° lines. Mark off the positions where the curves of the circles cross the ordinates. Thus a and b of drawing 4 are the same as a and b of drawing 1.

5   Sketch in freehand curves through the plot points. Note that the curves are ellipses.

6   Complete the drawing with firm smooth lines.

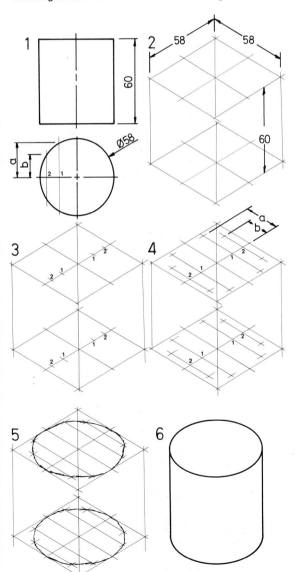

Exercises

1   Make the isometric drawing of the cylinder to a scale of 1:1 on a piece of A4 paper.

2   Make a scale 1:2 drawing of the reading lamp shade and stand. Use an A4 sheet of paper in the 'portrait' position – with its short edges horizontal. Choose your own dimensions for the decorative rectangles and circles of the base. Add a margin and a suitable title block to your drawing.

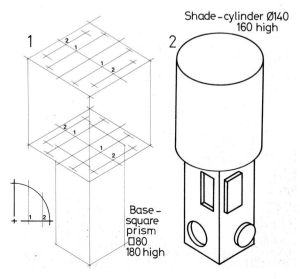

Shade – cylinder Ø140
160 high

Base –
square
prism
□80
180 high

Curves in isometric drawings, continued

The first drawing on this page describes the method of constructing an isometric drawing of a semi-elliptical curve. Note the method of drawing the lower sloping lines on the isometric drawing. The two 15 mm measurements must be made along the 30° axes of the drawing.

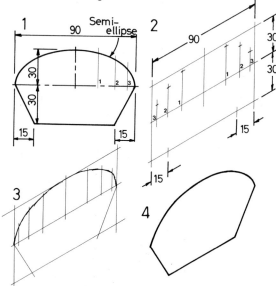

Four-arcs method of drawing ellipses

The ellipses produced on isometric drawings to show circular parts may be drawn by this geometrical method. It should be noted that the shapes produced by the four-arcs method are not true ellipses. This method should be confined to drawing small ellipses – say circles of Ø25 mm and smaller. Larger circles are best drawn in isometric drawings by plotting along ordinates as described on page 63.

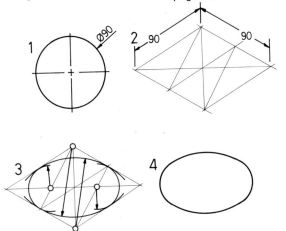

Exercises

The two following exercises can both be drawn on a single A4 sheet. Add a suitable title block to the sheet.

1   Make an accurate isometric drawing to a scale of 1:1 of the semi-elliptical sheet.

2   Make an accurate scale 1:1 isometric drawing of the turn-button. Take particular note of the method of drawing the upper curves. Drawings 2, 3 and 4 of the turn-button should be carefully followed.

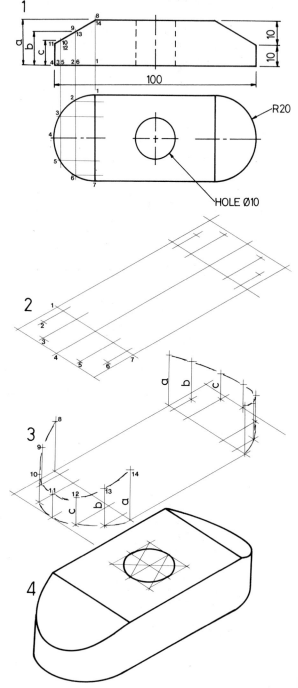

Freehand isometric drawing

The drawings on this page have been produced freehand without the aid of instruments. The production of pictorial freehand drawings along approximate isometric axes is a skill well worth developing.

Note the use of isometric grid papers as an aid to good freehand pictorial drawing.

Exercises

Copy the completed drawings shown on this page using dimensions of your own choice. Isometric grid paper, if available, may be used if thought suitable.

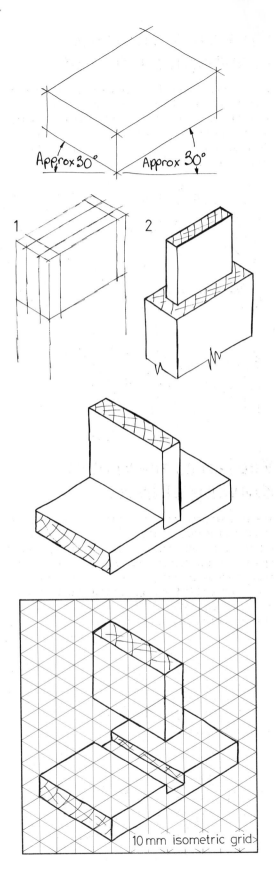

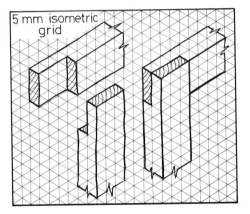

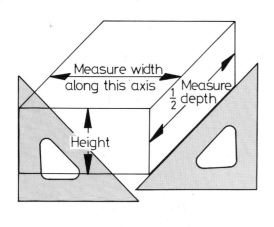

Measure width along this axis

Measure depth

$\frac{1}{2}$

Height

1

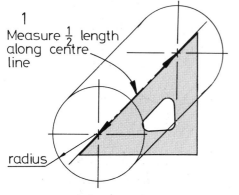

Measure $\frac{1}{2}$ length along centre line

radius

2

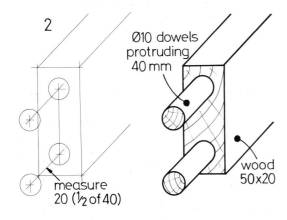

Ø10 dowels protruding 40 mm

measure 20 (½ of 40)

wood 50×20

3

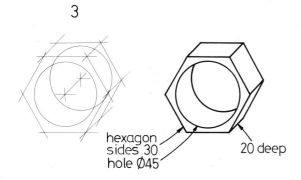

hexagon sides 30 hole Ø45

20 deep

4

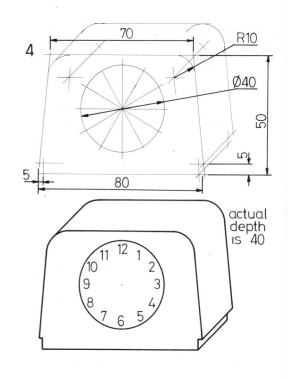

70

R10

Ø40

50

5

5

80

actual depth is 40

5

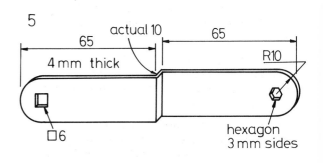

65

actual 10

65

4 mm thick

R10

□6

hexagon 3 mm sides

6

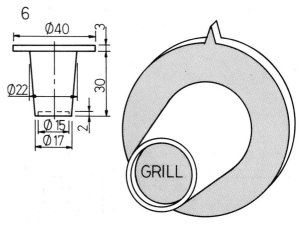

Ø40

3

Ø22

30

Ø15

Ø17

2

GRILL

Cabinet Drawing

The methods employed for producing cabinet drawings are shown in drawings on page 66 opposite. Cabinet drawings consist of front views with receding lines drawn at 45°. Lengths to half the scale used for the drawing are taken along the 45° lines.

Exercises

1   To any suitable dimensions make a cabinet drawing of a cylinder.

2   Make a scale 1:1 cabinet drawing of the dowel joint.

3   Construct a scale 1:1 cabinet drawing of the hexagonal prism with a hole.

4   Make a scale 1:1 cabinet drawing of the clock.

5   Make a scale 1:1 cabinet drawing of the brake spanner.

6   Make a scale 2:1 (twice full size) of the control switch from an electric cooker.

7   Figure 1 shows two views of a length of aluminium draw-pull. Make a scale 2:1 cabinet drawing of the length of draw-pull.

8   A 150 mm long scale rule is made from a length of the plastic section of equilateral triangular shape shown in Fig. 2. Make a scale 1:1 cabinet drawing of the rule. Ignore any calibrations.

9   Figure 3 shows two views of a square slab of aluminium with a circular recess turned in its surface. Construct a scale 1:1 cabinet drawing of the slab.

10   A plan of a circular door knob is given in Fig. 4. Construct a scale 2:1 cabinet drawing of the knob.

11   An electric shaver adaptor plug set into a wall socket is shown in Fig. 5. Draw a scale 2:1 cabinet drawing of the plug in its socket.

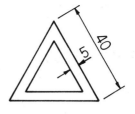

Fig. 2

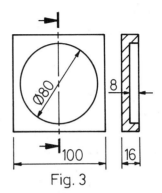

Fig. 3

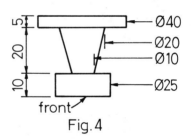

Fig. 4

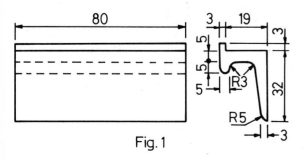

Fig. 1

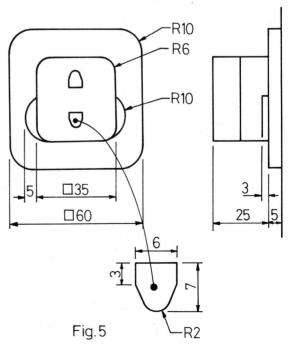

Fig. 5

Planometric Drawing

The methods employed for making planometric drawings are shown in the illustrations. Two angles of plan are suggested, the first drawn with a 30°, 60° set square, the second with a 45° set square. In the case of the plan at 45° it is suggested that verticals be shortened to $\frac{3}{4}$ of the size of the scale used for the drawing. This $\frac{3}{4}$ shortening is not critical. Some may prefer shortening to $\frac{2}{3}$ or even $\frac{1}{2}$, but full-size verticals give the drawing an appearance of being too tall.

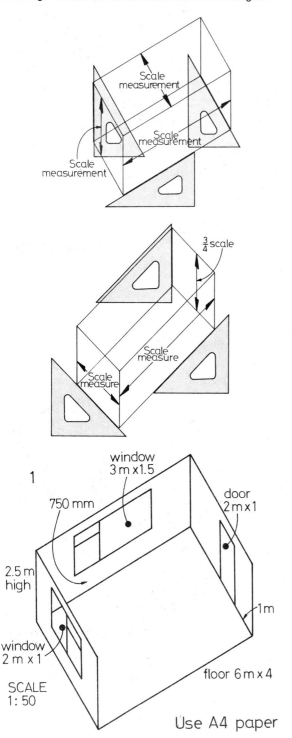

Scale measurement

Scale measurement

Scale measurement

$\frac{3}{4}$ scale

Scale measure

Scale measure

1

window
3 m x 1.5

door
2 m x 1

750 mm

2.5 m high

1m

window
2 m x 1

floor 6 m x 4

SCALE
1: 50

Use A4 paper

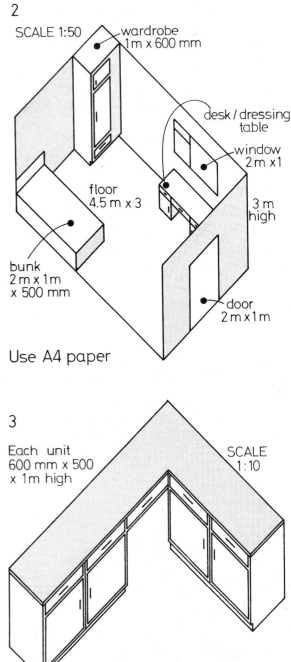

2

SCALE 1:50

wardrobe
1 m x 600 mm

desk / dressing table

window
2 m x 1

floor
4.5 m x 3

3 m high

bunk
2 m x 1m
x 500 mm

door
2 m x 1m

Use A4 paper

3

Each unit
600 mm x 500
x 1m high

SCALE
1:10

Use A4 paper

Exercises

1   Use A4 paper. Make the planometric drawing 1. Work to a scale of 1:50.

2   Use A4 paper. Add borders and a title. Make the planometric drawing 2 of a furnished room. Work to a scale of 1:50. The shaded parts of the drawing may be tinted by crayoning or colour wash. Use your discretion about dimensions not given.

3   Use A4 paper. Work to a scale of 1:10. Make the planometric drawing 3 of a set of kitchen units. The sizes not shown on the drawing can be estimated. Add borders and a title. Print your name, the scale and an appropriate title.

4   Use A4 paper. Make the planometric drawing 4 of a bath surround. Work to a scale of 1:10. Add borders and an appropriate title block.

5   Use A4 paper. Make the planometric drawing 5 to a scale of 1:50. Add a border and a title. Include in the title PLANOMETRIC DRAWING OF A GARAGE. Sizes not shown are left to your discretion.

6   Drawing 6 shows a planometric drawing of a house and garage in its garden plot. Using an appropriate scale make a planometric drawing of a similar house to your own designs. Use A4 paper. Add a border and an appropriate title.

5

Garage 7 m x 3 m
height at front 3 m
at rear 2.8 m

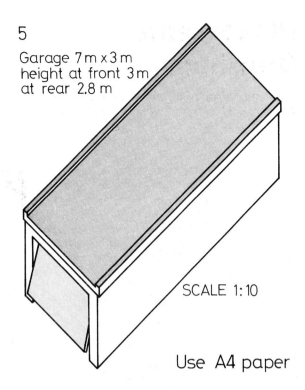

SCALE 1:10

Use A4 paper

4

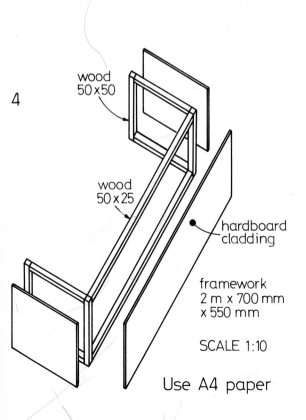

wood
50 x 50

wood
50 x 25

hardboard
cladding

framework
2 m x 700 mm
x 550 mm

SCALE 1:10

Use A4 paper

6

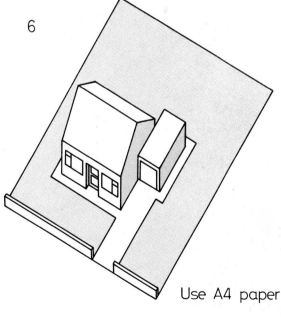

Use A4 paper

Orthographic Projection

Orthographic projection is the result of projecting the shapes of objects on to planes placed at right angles to each other.

Two methods of orthographic projection are practised in Britain. These are First Angle and Third Angle projections. The upper drawing shows that there are also a second and a fourth angle. The second and fourth angles cannot, however, be used for drawings.

The student is advised to practise making drawings in both projections. Each is of equal importance. The symbol for the projection to which a drawing has been made should be shown on the drawing sheet.

Note the following terms

Horizontal Plane Abbreviated to H.P.

Vertical Plane Abbreviated to V.P.

Front View The view as seen from that side of the object which has been selected as its front.

End View The view as seen from the end of the object.

Plan The view as seen from above the object.

Inverted Plan The view as seen from below the object.

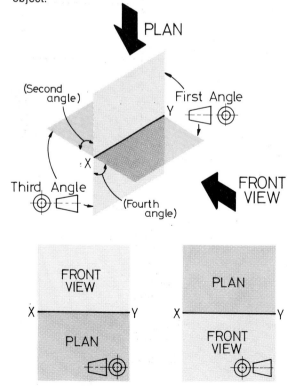

First Angle projection

Drawing 1 The end view as seen from the left of the front view is drawn to the right of the front view. The end view as seen from the right is drawn to the left of the front view. The plan is drawn below the front view and the inverted plan drawn above.

Third Angle projection

Drawing 2 The end view as seen from the left of the front view is drawn to the left of the front view. The end view as seen from the right is drawn to the right of the front view. The plan is drawn above the front view. The inverted plan is drawn below.

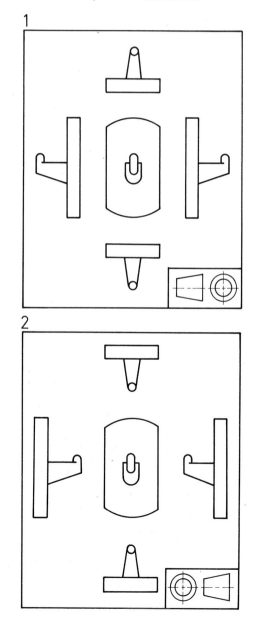

Symbols and Conventions

The more commonly used symbols and conventions for producing orthographic projection drawings are shown. These are based on the recommendations of British Standard 308 (Engineering Drawing Practice).

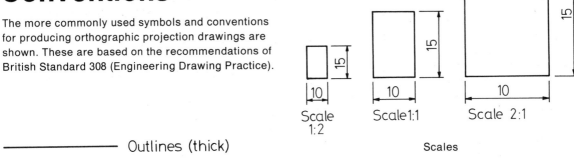

Scale 1:2   Scale 1:1   Scale 2:1

Scales

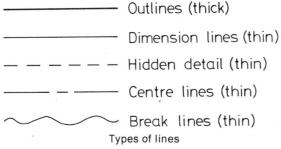

Outlines (thick)

Dimension lines (thin)

Hidden detail (thin)

Centre lines (thin)

Break lines (thin)

Types of lines

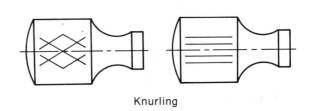

Knurling

ABCDEFGHIJKLMNOPQRSTUVWXYZ

abcdefghijklmnopqrstuvwxyz

1234567890

SCALE 1:2   PROJECTION   PIN

15 m   17.5 mm   4570   5 675   0.5

Letters and Figures

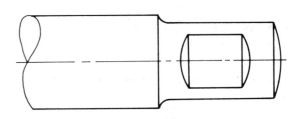

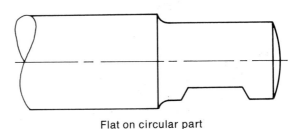

Flat on circular part

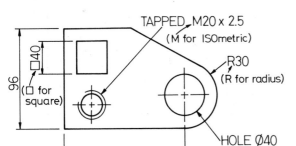

TAPPED M20 x 2.5
(M for ISOmetric)

R30
(R for radius)

□40
(□ for square)

96

120

HOLE ⌀40
(⌀ for diameter)

Dimensions

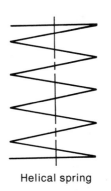

Helical spring

Freehand Orthographic Drawing

Orthographic projections drawn freehand may be made for many purposes in graphical communication. Some examples are given.

Drawing circles

This method of drawing circles freehand should be practised.

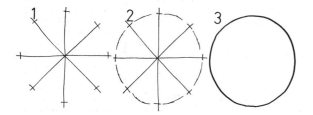

Preparation for finished drawings

The two drawings show typical freehand work in preparation for a final instruments drawing. After sketching outlines of the required three views, details of the views are completed freehand. The final instruments drawing is then commenced with the freehand work as a reference. The advantage of such freehand preparation drawings is that mistakes can be seen and amended before a final drawing is started.

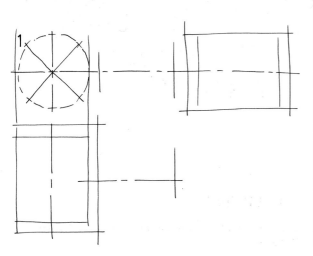

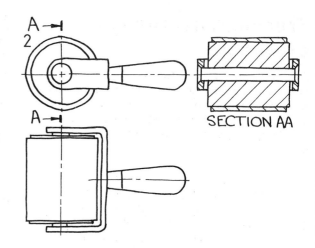

SECTION AA

Freehand drawing on grids

Two freehand preparations of orthographic drawings made on square grids are shown. The first is a single view drawing of a hand lens, the second of a socket spanner head showing three views. Final drawings with the aid of instruments are made from such drawings.

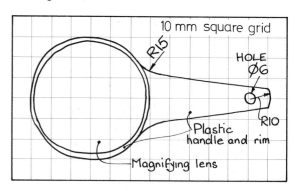

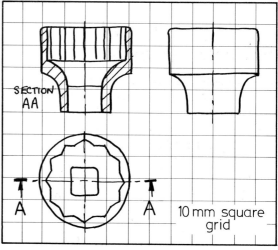

Freehand geometry drawings

Four examples of freehand drawings made as preparation for constructional geometry are shown. A simple freehand sketch of this nature will often assist in deciding which is the best construction to use. Colour may be added to freehand geometry drawings.

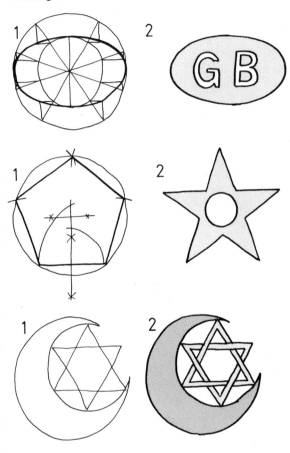

Drawing electrical circuits freehand

Freehand drawings of electrical circuits may be drawn as a preparation for drawing the circuit with instruments.

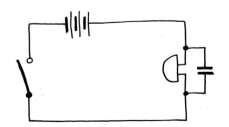

Freehand development drawings

The constructions of developments are described in the geometrical sections of Book 2. To assist in the construction of the shape of a development, a preliminary freehand drawing made on square grid paper may be drawn. Good freehand drawings may be tinted by colouring as shown in this example.

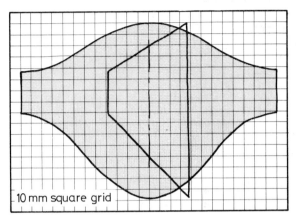

10 mm square grid

Freehand drawing of logograms

Drawings of graphics drawn freehand are a good preparation for final instrument drawings. Parts of such freehand drawings may be tinted with colours.

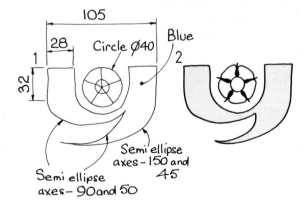

Exercises

Copy the freehand drawings on this page and page 72 opposite.

First Angle Orthographic Projection

A series of six drawings shows the method by which a television set is drawn in First Angle projection. Front view and plan are projected on to Vertical and Horizontal Planes. An end view is projected on to a second vertical plane. The planes are laid out flat to give the required projection with the views in correct relationship to each other.

Preparation for the finished drawing

A freehand drawing as a preparation for the working of the final drawing is always advisable. Difficulties of shape and of projection and the sizes needed for good spacing between views can be worked out on a freehand drawing before commencing the final drawing. When the spacing has been decided, border lines, title blocks and the outlines of the required views can be drawn with instruments on the correct size sheet of paper on which the finished projection is to be made.

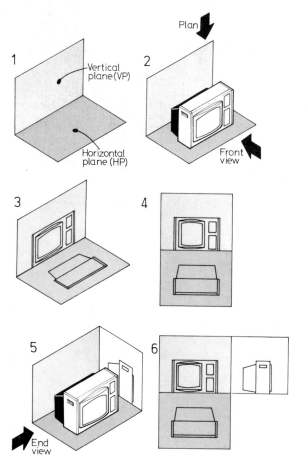

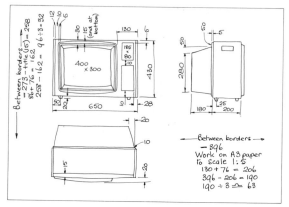

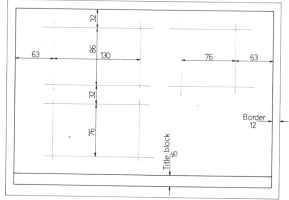

Television Set

The final first angle projection drawing of the television set is given. Note the following details.

1   Use A3 paper.

2   Border 12 mm from sheet edges.

3   Title block 15 mm high.

4   Neat lettering in title – 6 mm high.

5   Minimum information in title should be – name, form or group, date, scale and title of item drawn. Other information may be included in the title if thought necessary.

6   Projection symbol clearly displayed.

7   In this particular drawing only main dimensions have been included. It is usual to include all necessary dimensions.

8   Three views – front view, end view and plan, well spaced in accordance with the working shown on the freehand drawing on page 74 opposite.

Exercise

Copy the drawing of a TV set but place the end view to the left of the front view, noting the end view should then be with its front facing to the left.

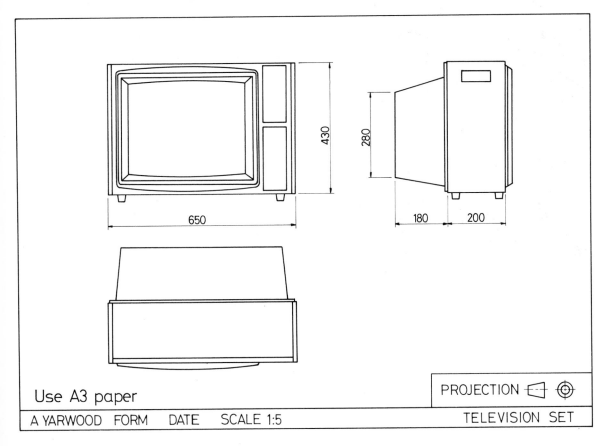

Use A3 paper

A YARWOOD   FORM   DATE   SCALE 1:5

PROJECTION

TELEVISION SET

Dowelling Jig

Details of a dowelling jig made from wood are given in a photograph and a dimensioned pictorial drawing.

Exercise

Copy the front view outline and the completed end view given. Complete the front view and add all necessary dimensions.

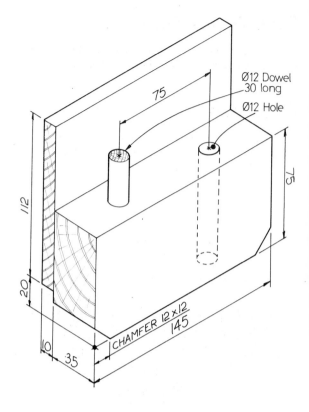

Ø12 Dowel
30 long

Ø12 Hole

75

75

112

20

10

35

CHAMFER 12×12

145

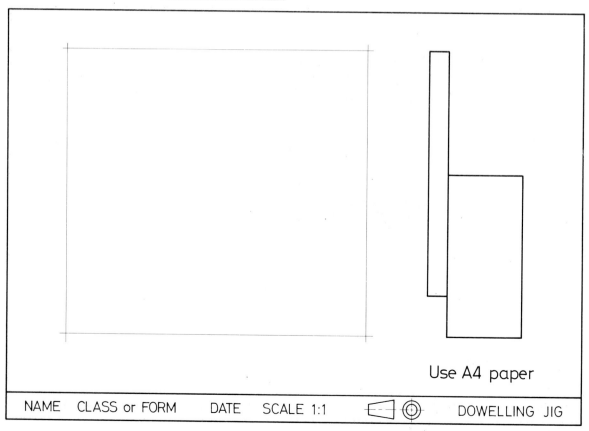

Use A4 paper

Disc Brake Pad

A photograph of a pair of disc brake pads taken from the front wheel braking system of a motor car is shown.

Exercise

Use an A4 sheet. Copy the given front view and outlines of the end view and the plan. Complete the end view and the plan. Add all necessary dimensions.

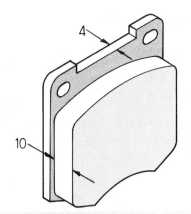

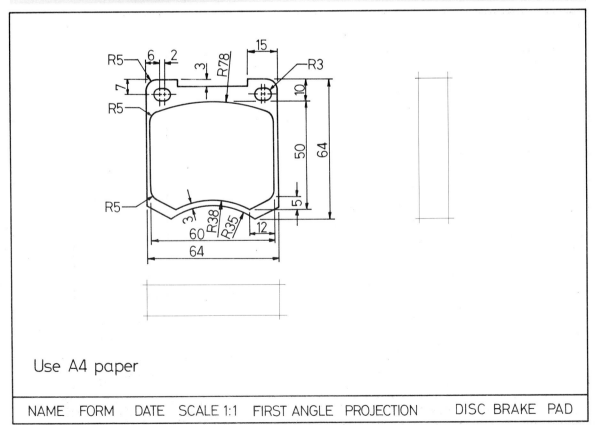

Use A4 paper

NAME	FORM	DATE	SCALE 1:1	FIRST ANGLE	PROJECTION	DISC BRAKE PAD

Coat Peg

A photograph and two dimensioned pictorial drawings of a coat peg are shown. The peg is made from plastic, and a metallic cap slides down over the rear of the peg after it has been screwed to a wall to cover the screw heads.

Exercise

Use A4 paper set upright in a 'portrait' position. Copy the given end view and the outlines of the front view and plan. Note that the end view is to the left of the front view. Complete the front view and plan. Add all necessary dimensions.

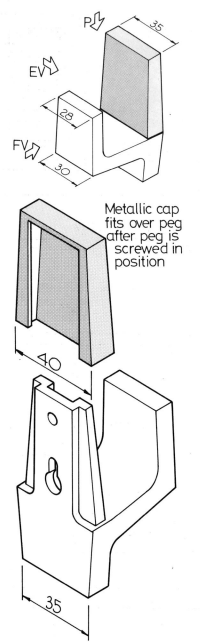

P⤵

EV⤴

FV⤵

35

28

30

Metallic cap fits over peg after peg is screwed in position

40

35

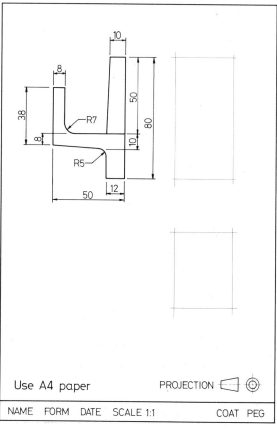

10

8

38

8

R7

R5

50

12

50

10

80

Use A4 paper   PROJECTION ◁ ⊙

| NAME | FORM | DATE | SCALE 1:1 | COAT PEG |

Third Angle Orthographic Projection

A series of three drawings shows the method by which a domestic weighing machine is drawn in Third Angle projection. Compare with First Angle projection on page 74. The completed drawing of the Third Angle projection of the weighing machine is also given.

Exercise

Use A3 paper. Work out on a freehand drawing the necessary spacing required between views to produce a good drawing layout. Copy the given drawing. Note its scale is 1:2. Sizes not given should be estimated.

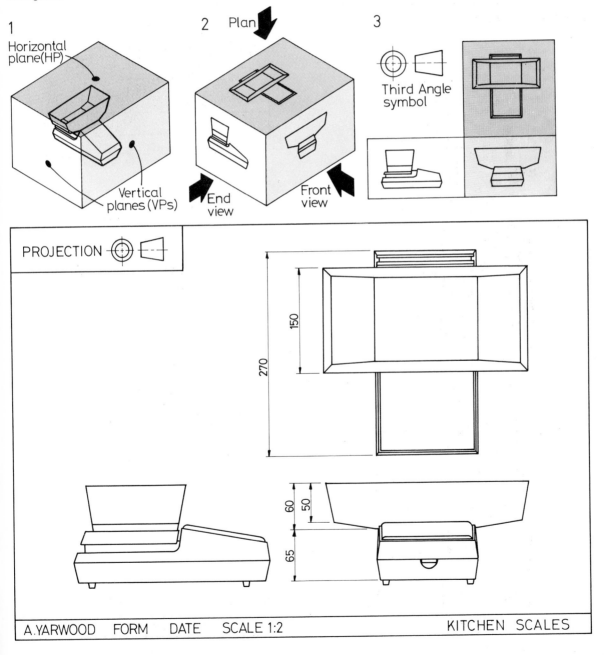

KITCHEN SCALES

A.YARWOOD   FORM   DATE   SCALE 1:2

Tooth Brush Rack

A photograph together with dimensioned drawings necessary for making a Third Angle projection of the rack are given.

Exercise

Use an A4 sheet. Copy the given plan and front and end view outlines. Complete the front and end views. Add necessary dimensions.

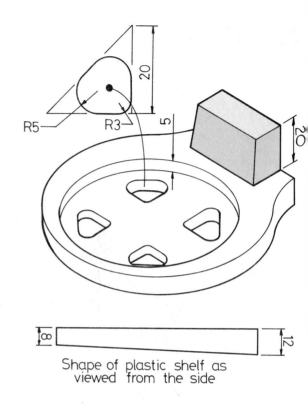

Shape of plastic shelf as viewed from the side

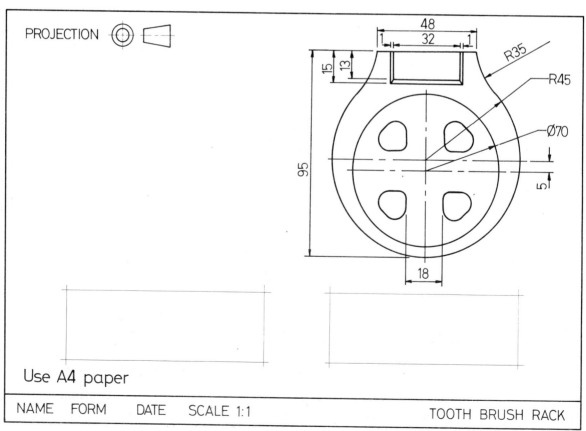

PROJECTION

Use A4 paper

| NAME | FORM | DATE | SCALE 1:1 | TOOTH BRUSH RACK |

Spacing Bracket

A photograph together with a partially completed Third Angle projection of a spacing bracket are given. This spacing bracket was taken from the radiator of a motor car. It was fitted to the car to attach the radiator to the engine body so as to leave sufficient space for the radiator fan to draw in air to cool the radiator.

Exercise

Use A3 paper. Copy the given front view and plan and the end view outline. Complete the end view. Add necessary dimensions.

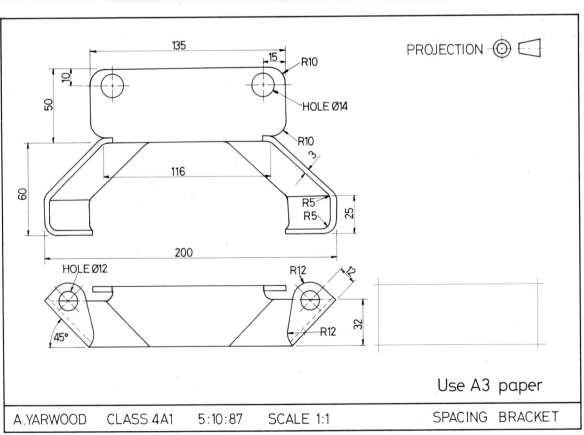

PROJECTION

Use A3 paper

A.YARWOOD   CLASS 4A1   5:10:87   SCALE 1:1   SPACING BRACKET

Transistor Set

A photograph and a dimensioned drawing of a transistor set are given.

Exercises

1 Make a neat freehand drawing of a front view, an end view and a plan of the transistor set. Arrows on the dimensioned drawing show the directions from which these three views should be taken. Work out the spacing necessary to draw a Third Angle projection of the three views to a scale of 1:1 on an A3 sheet of paper.

2 Use an A3 sheet. Draw accurately with instruments the three views in Third Angle projection which you have prepared in your freehand drawing. Add the three overall dimensions of the transistor set together with two diameter dimensions and the dimensions of the elliptical loud speaker.

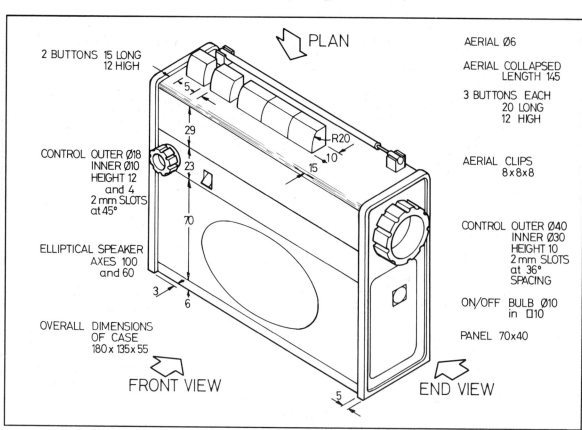

PLAN

2 BUTTONS 15 LONG
12 HIGH

CONTROL OUTER Ø18
INNER Ø10
HEIGHT 12
and 4
2 mm SLOTS
at 45°

ELLIPTICAL SPEAKER
AXES 100
and 60

OVERALL DIMENSIONS
OF CASE
180 x 135 x 55

FRONT VIEW

AERIAL Ø6

AERIAL COLLAPSED
LENGTH 145

3 BUTTONS EACH
20 LONG
12 HIGH

AERIAL CLIPS
8 x 8 x 8

CONTROL OUTER Ø40
INNER Ø30
HEIGHT 10
2 mm SLOTS
at 36°
SPACING

ON/OFF BULB Ø10
in □10

PANEL 70 x 40

END VIEW

R20
10
15
29
23
70
5
3
6
5

Electronic Calculator

A photograph and drawings of a small electronic desk calculator are given.

The end view should be as seen looking from the right-hand side of the given pictorial drawing.

Exercises

1   Make neat freehand drawings in Third Angle projection showing a front view, an end view and a plan of the calculator.

2   Make a freehand outline drawing to show the layout with dimensions between views to enable the three views to be drawn to a scale of 1:1 on a sheet of A3 size paper.

3   With the aid of the two freehand drawings make a scale 1:1 Third Angle projection of the calculator on an A3 sheet. Add a title block and include the following dimensions:
(a)   Overall length, width and height
(b)   The sizes of the keys
(c)   The sizes of the numbers window

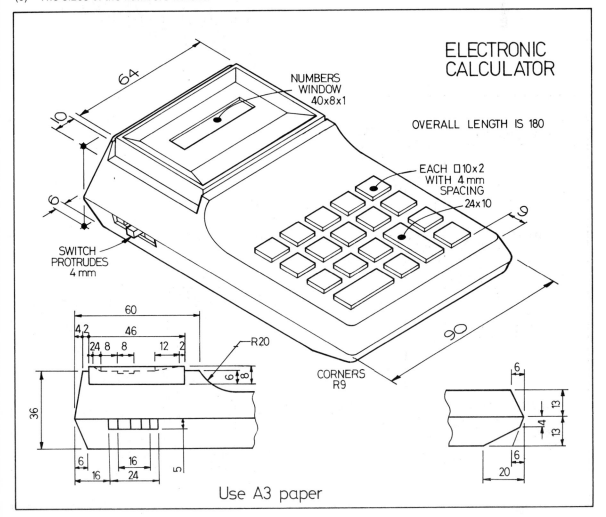

ELECTRONIC
CALCULATOR

NUMBERS WINDOW 40x8x1

OVERALL LENGTH IS 180

EACH □10x2 WITH 4 mm SPACING 24x10

SWITCH PROTRUDES 4 mm

64

R20

CORNERS R9

90

Use A3 paper

Cassette Player

A photograph and freehand drawings of a cassette player are given. The drawings contain sufficient dimensions to enable the cassette player to be drawn in orthographic projection. Any dimensions not given are left to your own judgement.
Use Third Angle projection.

Exercises

1 Make a neat Third Angle freehand drawing showing
(a) a front view
(b) an end view as seen in the direction of EV1
(c) an end view as seen in the direction of EV2
(d) a plan.

2 Make a freehand layout drawing to show the spacing between the four views necessary to achieve a good layout.

3 Use A2 paper. Work to a scale of 1:1. Draw accurately, with instruments, the four views from your freehand drawing. Add at least six dimensions which are as varied as possible.

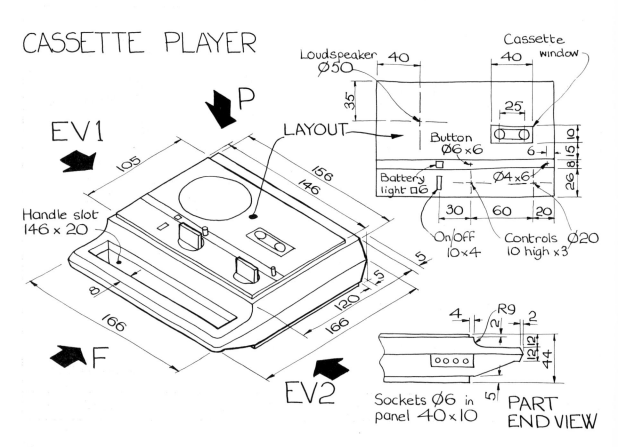

CASSETTE PLAYER

EV1

P

Handle slot
146 × 20

Loudspeaker
Ø50

LAYOUT

Cassette window

Button
Ø6 × 6

Battery
light □6

Ø4 × 6

On/off
10 × 4

Controls Ø20
10 high × 3

F

EV2

Sockets Ø6 in
panel 40 × 10

R9

PART
END VIEW

Sections

Drawing sections is a very important part of orthographic projection. Sections show the shapes of those parts of an object which cannot clearly be seen from outside the object.

To draw a section you must imagine that the object has been cut along the line through which you require to draw the sectional shape. When the section has been drawn, its cut surfaces should be 'hatched' with thin 45° lines at intervals across the cut surfaces.

Half sections

If the object is symmetrical about an axis, it is often possible to describe the internal shape of the object by drawing a half section. Note that it is unusual to include hidden detail in a half section. A little thought will show that it is usually unnecessary to do so.

Note Sectional drawings should include:

1 A section plane line showing the edge of the section plane in position cutting the object.

2 45° hatching of the sectional parts set at about 4 mm apart.

3 A label near the sectional view stating to which section plane the section refers.

Exercise

Make a neat drawing with instruments of the plant pot drawing given on this page on A4 paper. Add the correct projection symbol to your drawing.

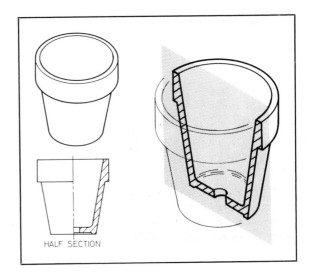

HALF SECTION

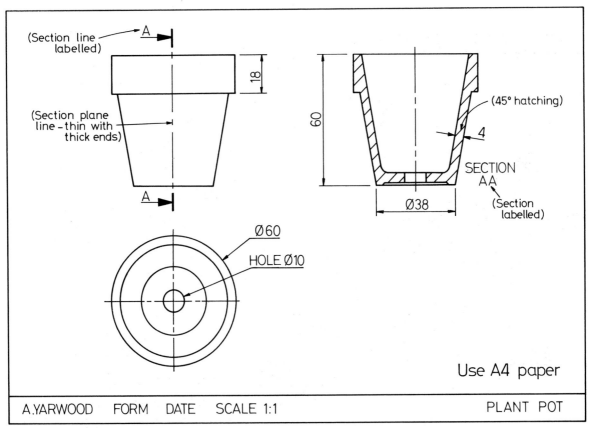

(Section line labelled)

A

18

(Section plane line – thin with thick ends)

A

(45° hatching)

60

4

Ø38

SECTION AA

(Section labelled)

Ø60

HOLE Ø10

Use A4 paper

A.YARWOOD   FORM   DATE   SCALE 1:1   PLANT POT

Book Trough

Dimensions may be taken either from the pictorial drawing or from the end view of the orthographic drawing.

Exercise

Make an accurate scale 1:2 copy of the given orthographic drawing in First Angle projection on A3 paper. Complete the front view and plan and add necessary dimensions.

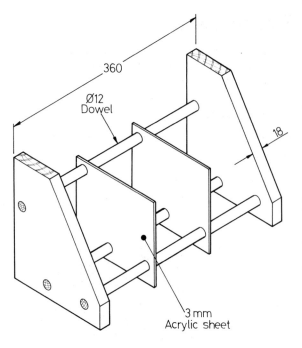

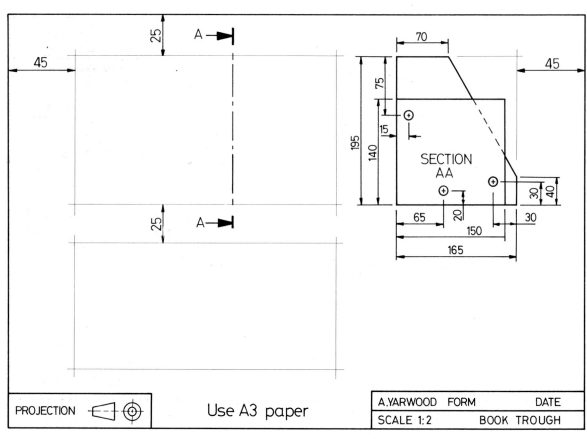

Brake Shoe

A photograph of a pair of drum brake shoes from a motor car, and a dimensioned front view showing how they fit into the brake drum case, are given.

Exercise

Use A4 paper. Do not copy the given projection, but work in Third Angle.

In Third Angle, draw the two given views and complete the plan. Include all dimensions and add dimensions to the plan where necessary.

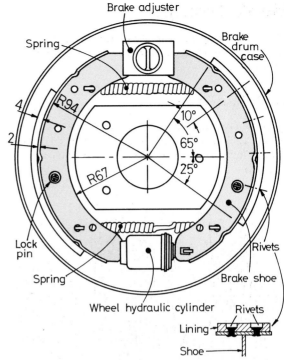

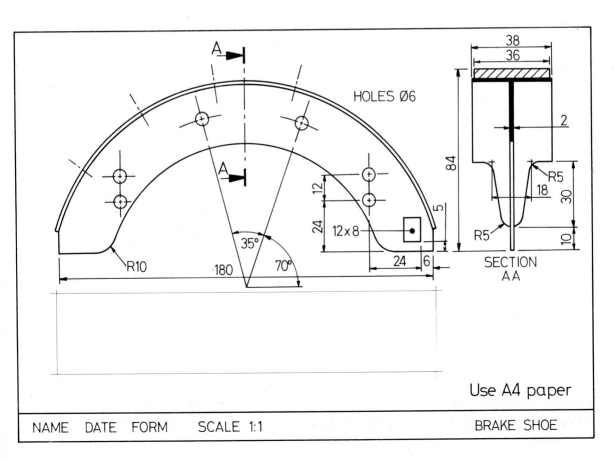

HOLES Ø6

35° 180 70°

R10

12 24 12 x 8 24 6 5

38 36 2 R5 18 30 10 84

SECTION A A

Use A4 paper

NAME   DATE   FORM   SCALE 1:1   BRAKE SHOE

Pipe Bending Die

A photograph shows a 19 mm pipe bending die fitted to a bench pipe bending machine. The drawing shows a front view and plan in First Angle projection of a 16 mm die for the same machine.

Exercise

Copy the given drawing on a sheet of A3 paper and complete the end view. Add dimensions as thought necessary.

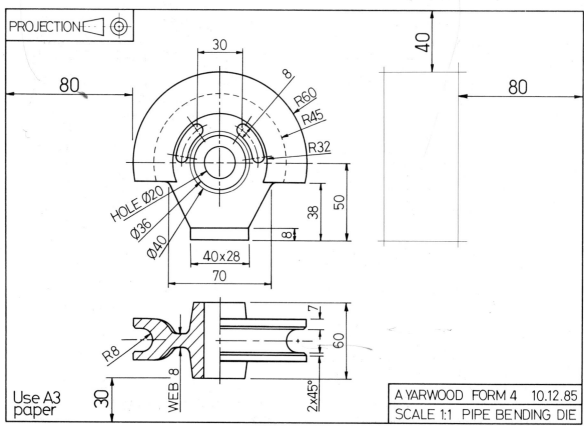

PROJECTION

30
8
R60
R45
R32
80
HOLE Ø20
Ø36
Ø40
50
38
8
40×28
70
40
80

R8
WEB 8
30
7
60
2×45°

Use A3 paper

A YARWOOD   FORM 4   10.12.85

SCALE 1:1   PIPE BENDING DIE

Screwed Fastenings

All the threaded screws shown are drawn in First Angle projection. The conventional method of drawing screwed parts of any type is to draw thin lines at approximately thread depth from the screw outlines. Screw threads end with thick lines drawn across the screw with thin run-off lines at 30°. Threads are shown on end views by thin broken lines.

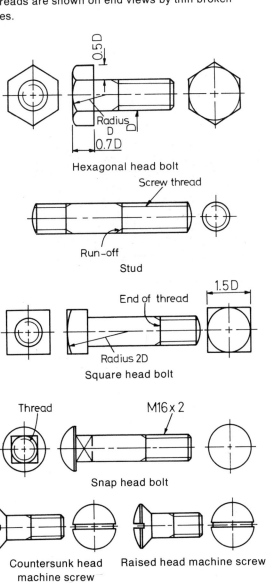

0.5 D

Radius D

0.7 D

Hexagonal head bolt

Screw thread

Run-off

Stud

1.5 D

End of thread

Radius 2D

Square head bolt

Thread   M16 x 2

Snap head bolt

Countersunk head machine screw   **Raised head machine screw**

Ø2D

Pan head machine screw   **Cheese head machine screw**

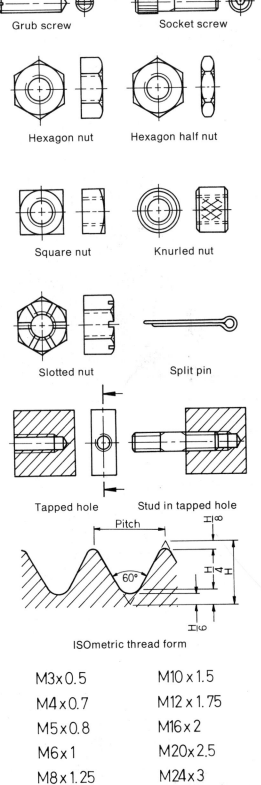

Grub screw   Socket screw

Hexagon nut   Hexagon half nut

Square nut   Knurled nut

Slotted nut   Split pin

Tapped hole   Stud in tapped hole

Pitch

60°

ISOmetric thread form

M3 x 0.5	M10 x 1.5
M4 x 0.7	M12 x 1.75
M5 x 0.8	M16 x 2
M6 x 1	M20 x 2.5
M8 x 1.25	M24 x 3

Dimensions of some standard ISOmetric screw threads

Film Drying Clip

The photograph on this page shows a film drying clip such as is used for hanging wet film for drying after it has been washed after developing. The exploded pictorial drawing on page 91 opposite is a large-size drawing of the same clip. Exploded drawings of this type are made to show how objects made from several parts are fitted together.

Exercise

Use a sheet of A3 paper. Fix the sheet in 'portrait' position on your drawing board. Draw a border line and add a title block. Print clearly the following information in the title block:

(a)   Your name
(b)   Your form or group number
(c)   The scale – 2:1
(d)   The date
(e)   The title – FILM DRYING CLIP

Before commencing the finished drawing make a neat freehand sketch showing a front view and a sectional end view of the clip. Add to this freehand drawing the spacing sizes between the views necessary to obtain a good layout.

Draw with instruments the following two views of the assembled clip. Work in Third Angle projection to a scale of 2:1.

1   A front view as seen from the same direction in which the photograph was taken.

2   A sectional end view to whichever side of the front view you choose. The sectional plane should be central to the front view.

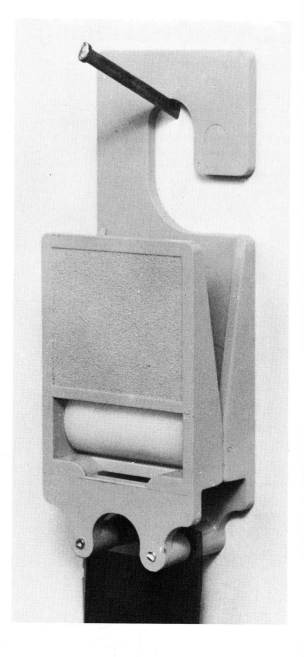

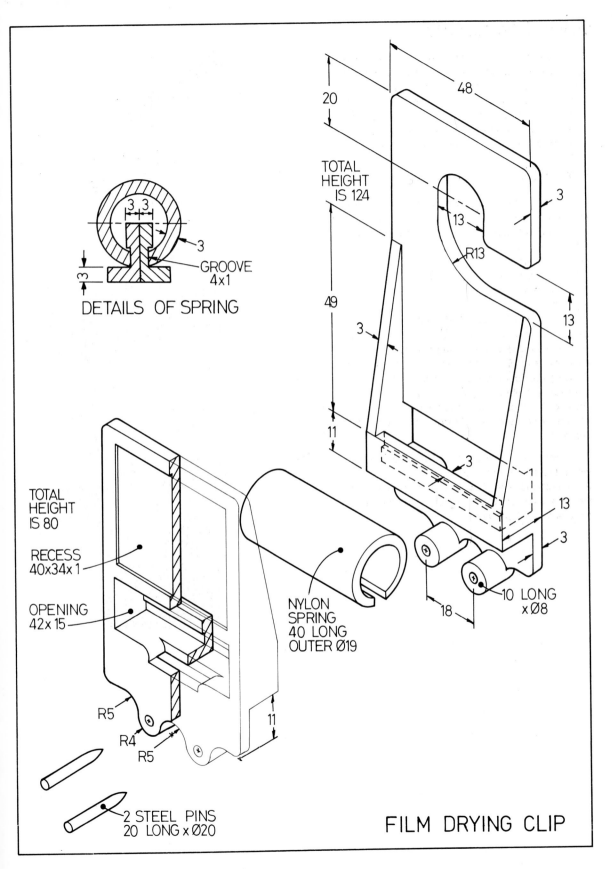

DETAILS OF SPRING

3 3

3

3

GROOVE
4x1

TOTAL
HEIGHT
IS 124

20

48

3

13

R13

49

3

13

11

3

13

3

TOTAL
HEIGHT
IS 80

RECESS
40x34x1

OPENING
42x15

NYLON
SPRING
40 LONG
OUTER Ø19

18

10 LONG
x Ø8

R5

R4

R5

11

2 STEEL PINS
20 LONG x Ø20

FILM DRYING CLIP

Retort Clamp

A photograph and a partly completed First Angle projection of a retort clamp are given.

Exercise

Use A3 paper. Do not copy the given drawing but work instead in Third Angle projection. Draw the given views in Third Angle and complete the front view. No further dimensions need be added.

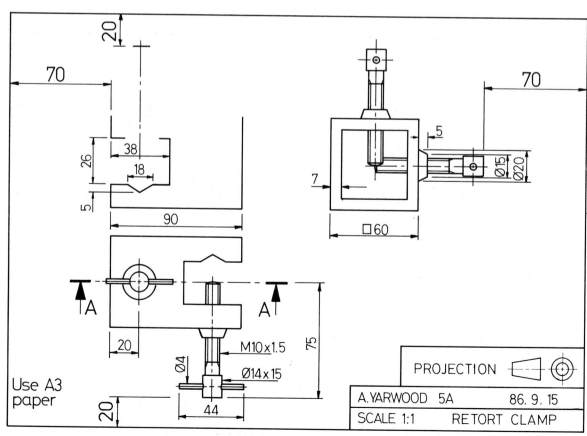

Use A3 paper

PROJECTION

A. YARWOOD 5A   86. 9. 15

SCALE 1:1   RETORT CLAMP

Lens Holder

A photograph and dimensioned isometric drawings of a lens holder from a slide projector are given.

Exercises

1 Work in First Angle projection. Make neat freehand sketches of the front view as seen in the direction of arrow F, an end view to the left of the front view and a plan. Work out layout spacing on this drawing.

2 Use A3 paper. Draw borders and a suitable title block. Complete the title using 6 mm high printing.

3 Draw accurately with instruments, in First Angle projection, the three views from your freehand drawing. Add dimensions as follows:
(a) Overall height
(b) Overall length
(c) A radius
(d) A diameter
(e) A screw thread

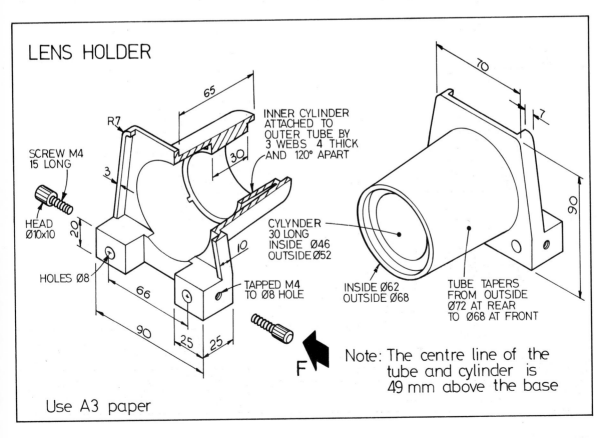

LENS HOLDER

65

R7

SCREW M4
15 LONG

HEAD
Ø10x10

3

INNER CYLINDER
ATTACHED TO
OUTER TUBE BY
3 WEBS 4 THICK
AND 120° APART

30

20

CYLYNDER
30 LONG
INSIDE Ø46
OUTSIDE Ø52

10

HOLES Ø8

66

TAPPED M4
TO Ø8 HOLE

90

25

25

F

INSIDE Ø62
OUTSIDE Ø68

TUBE TAPERS
FROM OUTSIDE
Ø72 AT REAR
TO Ø68 AT FRONT

70

7

90

Note: The centre line of the
tube and cylinder is
49 mm above the base

Use A3 paper

CROSS FEED MECHANISM FOR SHAPING MACHINE

PROJECTION

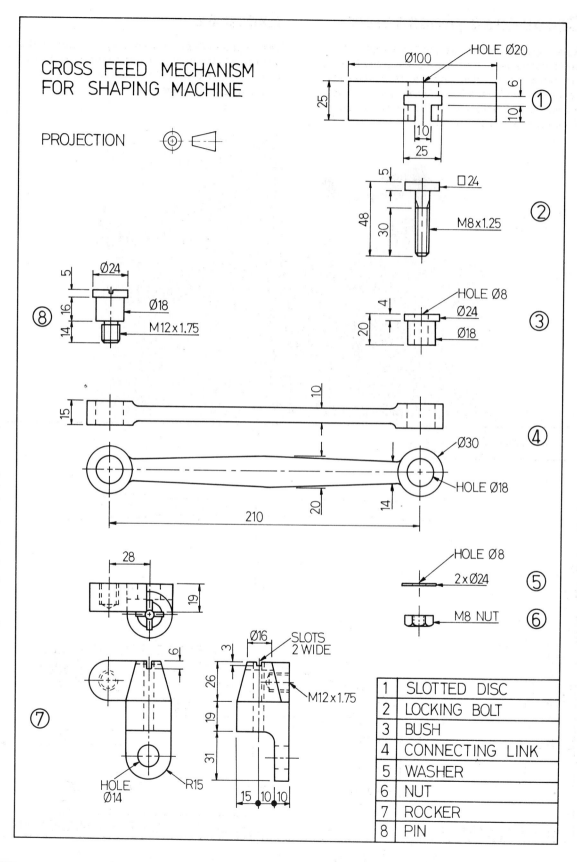

1	SLOTTED DISC
2	LOCKING BOLT
3	BUSH
4	CONNECTING LINK
5	WASHER
6	NUT
7	ROCKER
8	PIN